T0140720

Numerical Analysis
of
Nonlinear Partial Differential-Algebraic Equations:

A Coupled and an Abstract Systems Approach

Inaugural-Dissertation

zur

Erlangung des Doktorgrades

der Mathematisch-Naturwissenschaftlichen Fakultät

der Universität zu Köln

vorgelegt von

Michael Matthes

aus Basel

Köln, 2012

Berichterstatter/in: Prof. Dr. Caren Tischendorf
Prof. Dr. Tassilo Küpper

Tag der mündlichen Prüfung: 05.11.2012

Bibliografische Information der Deutschen Nationalbibliothek

Die Deutsche Nationalbibliothek verzeichnet diese Publikation in der
Deutschen Nationalbibliografie; detaillierte bibliografische Daten sind
im Internet über http://dnb.d-nb.de abrufbar.

©Copyright Logos Verlag Berlin GmbH 2012
Alle Rechte vorbehalten.

ISBN 978-3-8325-3278-9

Logos Verlag Berlin GmbH
Comeniushof, Gubener Str. 47,
10243 Berlin
Tel.: +49 (0)30 42 85 10 90
Fax: +49 (0)30 42 85 10 92
INTERNET: http://www.logos-verlag.de

Kurzzusammenfassung

Verschiedene mathematische Modelle in zahlreichen Anwendungsgebieten resultieren in Gleichungssystemen Differential-Algebraischer Gleichungen (DAEs) und partieller Differentialgleichungen. Diese Systeme werden auch partielle oder abstrakte Differential-Algebraische Gleichungen genannt (ADAEs). Sie werden gewöhnlich durch die Linienmethode diskretisiert und das halb-diskretisierte System stellt eine DAE dar. Eine systematische mathematische Behandlung nichtlinearer ADAEs ist zurzeit noch im Anfangsstadium.

Wir präsentieren zwei Ansätze, um nichlineare ADAEs zu behandeln. Zum einen untersuchen wir sie im Hinblick auf Lösbarkeit und Eindeutigkeit von Lösungen und der Konvergenz von Lösungen des halb-diskretisierten Systems zur Lösung des Ausgangssystems. Zum anderen studieren wir die Sensitivität der Lösung unter Störungen auf der rechten Seite und im Startwert.

Der erste Ansatz erweitert einen Ansatz von Tischendorf für die Behandlung einer speziellen Klasse linearer ADAEs auf den nichtlinearen Fall. Er basiert auf dem Galerkinansatz und der Theorie monotoner Operatoren für Evolutionsgleichungen. Wir zeigen eindeutige Lösbarkeit der ADAE und starke Konvergenz der Galerkinlösungen. Weiterhin beweisen wir, dass diese Klasse von ADAEs Störungsindex 1 und höchstens ADAE-Index 1 hat.

Im zweiten Ansatz formulieren wir zwei Prototypen gekoppelter Systeme, einen elliptischen und einen parabolischen, indem wir eine semi-explizite DAE an eine unendlich dimensionale algebraische Operatorgleichung oder an eine Evolutionsgleichung koppeln. Für beide Prototypen zeigen wir eindeutige Lösbarkeit, starke Konvergenz der Galerkinlösungen und ein Störungsindex-1-Resultat. Beide Prototypen finden Anwendung bei konkreten gekoppelten Systemen in der Schaltungssimulation. In diesem Zusammenhang zeigen wir zusätzlich globale Lösbarkeit für die nichtlinearen Gleichungen der Modifizierten Knotenanalyse unter passenden topologischen Voraussetzungen.

Abstract

Various mathematical models in many application areas give rise to systems of partial differential equations and differential-algebraic equations (DAEs). These systems are called partial or abstract differential-algebraic equations (ADAEs). Being usually discretized by the method of lines the semi-discretized system yields in general a DAE. A substantial mathematical treatment of nonlinear ADAEs is still at an initial stage.

We present two approaches treating nonlinear ADAEs. We investigate them with regard to the solvability and uniqueness of solutions and the convergence of solutions of semi-dicretized systems to the original solution. Furthermore we study the sensitivity of a solution with regard to perturbations on the right hand side and in the initial value.

The first approach represents an extension of an approach by Tischendorf for the treatment of a specific class of linear ADAEs to the nonlinear case. It is based on the Galerkin approach and the theory of monotone operators for evolution equations. We prove unique solvability of the ADAE and strong convergence of the Galerkin solutions. Furthermore we prove that this class of ADAEs has Perturbation Index 1 and at most ADAE Index 1.

In the second approach we formulate two prototypes of coupled systems, an elliptic and a parabolic one. Here a semi-explicit DAE is coupled to an infinite dimensional algebraic operator equation or an evolution equation. For both prototypes we prove unique solvability, strong convergence of Galerkin solutions and a Perturbation Index 1 result. Both prototypes are applied to concrete coupled systems in circuit simulation. In this context we also prove a global solvability result for the nonlinear equations of the Modified Nodal Analysis under suitable topological assumptions.

In memory of my father

Acknowledgement

First, I owe my gratitude to Prof. Dr. Caren Tischendorf who encouraged me to delve into the world of abstract differential-algebraic equations. I would like to thank her, first and foremost, for the guidance and support she offered when pursuing new ideas. Furthermore I am deeply grateful for her help and patience in setting up the project "Numerische Analysis Abstrakter Differential-Algebraischer Gleichungen" funded by the Deutsche Forschungsgemeinschaft. It helped me a lot in the past months to finish this thesis.

Furthermore I would like to thank Prof. Dr. Tassilo Küpper for his decision to serve as a referee of this thesis, Prof. Dr. Rainer Schrader for aggreeing to be the chairperson of my doctoral committee and Dr. Roman Wienands to be a committee member.

I am deeply indebted to my many colleagues at the University of Cologne, in particular Dr. Sascha Baumanns, Lennart Jansen, Dr. Monica Selva Soto and Dr. Michael Menrath – not only for all the discussions on mathematical topics but also for being friends in the past years.

Furthermore I would like to thank Dr. Olivier Verdier who was the first one with whom I discussed my ideas on abstract differential-algebraic equations, Prof. Dr. Etienne Emmrich for letting me participate in the 2nd Spring School about "Analytical and Numerical Aspects of Evolution Equations", where I learned a lot about monotone operator theory, and Dr. Andreas Bartel who by sending me a copy of his doctoral thesis brought coupled circuit simulation with thermal effects to my attention.

I also like to thank all my friends at the University of Cologne and the Fraunhofer Institute SCAI for making lunch and coffee breaks most enjoyable, especially Sven Ulbrich, Konstantin Glombek, Leonid Torgovitski, Hella Timmermann, Stefan Müller, Dr. Matthias Rettenmeier, Josua Lidzba and Malte Förster.

Special thanks go to Yvonne Havertz and my sister Sabine for proofreading this thesis.

Finally, my deepest gratitude goes to my mother and my family for their love and support throughout my life. Without their trust and encouragement this thesis would have not been possible.

Michael Matthes

Cologne, November 8, 2012.

Contents

1. Introduction   **1**

2. A general setting for ADAEs   **5**
   2.1. Function spaces and norms 5
   2.2. A general form for ADAEs 9
   2.3. Conclusion 20

3. Solvability of nonlinear DAEs with monotonicity properties   **23**
   3.1. Preliminaries 24
   3.2. Solvability results for DAEs 30
   3.3. Application in circuit simulation 42
   3.4. Conclusion 53

4. Nonlinear ADAEs with monotone operators   **55**
   4.1. Structural assumptions 56
   4.2. The Galerkin equations 69
   4.3. Unique solvability 82
   4.4. Dependence on the data 91
   4.5. Conclusion 98

5. Coupled Systems   **101**
   5.1. An elliptic prototype 102
   5.2. Application of the elliptic prototype in circuit simulation 112
   5.3. A parabolic prototype 119
   5.4. Application of the parabolic prototype in circuit simulation 137
   5.5. Conclusion 149

6. Conclusion and Outlook   **151**

A. Appendix   **153**
   A.1. Projectors 153
   A.2. Theorem of Carathéodory 154
   A.3. Basics for Evolution Equations 155
   A.4. Theorem of Browder-Minty 160
   A.5. Theorem of Arzelà-Ascoli 162

Bibliography   165

Notation   173

Index   177

1. Introduction

Numerous mathematical models in science and engineering give rise to systems comprising partial differential equations (PDEs) and differential-algebraic equations (DAEs). These systems are called partial differential-algebraic equations (PDAEs) and occur frequently in application areas such as electric circuit simulation, flexible multibody systems, gas or water distribution network simulation or chemical engineering, see [CM96b, Sim00, Tis04]. Loosely speaking DAEs differ from explicit ordinary differential equations (ODEs) as the dynamics of the system are restricted to some algebraic constraints. Furthermore considering the temporal dynamics of PDE systems, constraints may occur when e.g. considering mixed systems involving elliptic and parabolic PDEs.

Starting in the 1980s systems of linear PDEs were studied by applying the method of lines. This means the complex system, depending on spatial variables and the time, is first discretized in all spatial variables. The resulting system of equations depends on time only and yields an ODE or a DAE. This was a major reason for DAEs becoming a subject of systemcatic research and it is also a reason why these systems are often called PDAEs. The term PDAE is frequently used when studying the semi-discretized system which is in fact a DAE. Although the term PDAE is more common in literature we will use the term abstract differential-algebraic equation (ADAE) to ensure that the non-discretized system is considered. Most of the mathematical analysis has been done for discretized PDAEs on the basis of the theory of DAEs which was well developed in the past thirty years, see the standard textbooks [GM86, HLR89, BCP96, KM06, Ria08, LMT13]. However, the theoretical treatment of especially nonlinear ADAEs is still at an initial stage.

A useful tool for classifying DAEs is its index which gives an insight into the solution and perturbation behavior. There are several index concepts available for DAEs. Regarding solvability they have proven fruitful for analyzing the structure of the DAE and thus identifying the inherent dynamics and all (hidden) constraints. So initial values cannot be chosen freely as it is the case for ODEs. For higher index DAEs not only integration but also differentiation problems occur. Furthermore DAEs of higher index represent ill-posed problems. This means that small perturbations on the right hand side may lead to large differences in the solution. This dependence is measured by the Perturbation Index and is very important for the numerical treatment of DAEs, see [HLR89, LMT13].

A similar tool would be desirable for ADAEs. The definition and determination of indexes for linear ADAEs has received attention in recent literature, see [CM96b, LSEL99, LMT01, Tis04, RA05, Rei06]. A comprehensive theory for existence and uniqueness of solutions to (nonlinear) ADAEs does not exist. Even in the case of standard PDEs, where classifications in terms of elliptic, parabolic or hyperbolic PDEs exist, different approaches are used. They are usually tailored to the particular properties of the investigated equation. For ADAEs the situation becomes even more complex because various types of equations are mixed.

In this thesis we are interested in a systematic treatment of nonlinear ADAEs. We derive certain classes of ADAEs which we are going to investigate with regard to the following guiding questions.

1. What conditions must be met for an ADAE to be solvable? When is the solution unique?

2. How does the solution change if the system is perturbed? Are there index criteria – similar to DAEs – which describe the perturbation behavior?

3. How should an ADAE be discretized? Is the solution of the semi-discretized system unique and does it converge to the exact solution?

While the first question is a core question in mathematics, the second one aims at suitable perturbation estimates. The second and the last question are closely linked. It is known that using different discretization schemes the semi-discretization of the ADAE may act like a deregularization (increasing the Perturbation Index) or regularization (decreasing the Perturbation Index), see [Gün00]. So the determination of a perturbation estimate for the ADAE is important to predict the perturbation behavior to be expected from a corresponding discretized system. Furthermore it should be stressed that the convergence of solutions of the semi-discretized ADAE to a solution of the original ADAE is merely investigated in literature where numerical analysis is mainly concerned with the semi-discretized ADAE itself and its discretization in time.

We concentrate on two approaches towards the treatment of ADAEs in the sense of the guiding questions above. The first one is based on the solvability theory for evolution equations. Common approaches are the semigroup approach and the Galerkin approach, see [GGZ74, Zei90b, Rou05]. For the semigroup approach solvability results for ADAEs – in this context also called degenerate differential equations – exist, mainly for the linear case, see [FY99], but also for certain classes of nonlinear ADAEs, see [FR99, RK04]. The Galerkin approach is preferable in our context as it naturally provides a numerical solution procedure. It is based on the theory of monotone operators in Banach spaces which will be an essential tool for proving our results. The Galerkin approach for evolution equations has been extended to a specific class of linear ADAEs in [Tis04]. There a solution is found that is only differentiable in the sense of generalized derivatives in the context of evolution triples. We will extend this result to the nonlinear case. A major problem that occurs here is the solvability of the nonlinear Galerkin equations because the standard solvability theory for DAEs cannot be applied due to lacking smoothness properties.

The second approach is tailored to treat so-called coupled systems. They arise especially in circuit simulation and have gained increasing significance in the last ten to fifteen years. Ever more increasing demands on high performance chips result in higher complexity, package densities and operating frequencies of integrated circuits. The well-established standard approach of describing the behavior of the circuit by the equations of the Modified Nodal Analysis (MNA), see [CL75, DK84, CDK87], is not sufficient anymore for capturing all physical effects. Electromagnetic effects or heating effects, stemming from the surrounding circuitry, for example, or more accurate switching behavior of semiconductors cannot be neglected anymore when simulating correct physical behavior. So the MNA equations, yielding a DAE, are usually complemented by a suitable system of PDEs describing these additional effects or elements. The MNA equations and the PDE system interchange information via certain coupling terms, additional coupling equations, certain variables serving as input for boundary conditions of the PDE system or even more general parametric coupling. The resulting system is in general very complex and nonlinear, see e.g. [Gün01, ABGT03, Tis04, Bar04, ABG05, Cul09, ABG10, Sch11, Bau12].

As a new systematic approach towards coupled systems we present two general prototype systems. On the one hand a semi-explicit DAE is coupled to an operator equation (elliptic prototype) and on the other hand it is coupled to an evolution equation (parabolic prototype). In both cases we derive unique solvability results and show strong convergence of the solutions to the corresponding Galerkin equations. Furthermore we show that these systems have Perturbation Index 1.

These prototypes can also be applied to coupled systems in circuit simulation. Here the MNA equations are coupled to the Laplace equations for simulating a specific resistor and the heat equation for simulating additional heating effects. In both cases a decoupling of the MNA equations is necessary. With this decoupling and adapted passivity conditions for the basic elements of the circuit we also derive a global solvability result for the MNA equations under suitable topological conditions.

This thesis is organized in the following way. After some preliminaries on function spaces and norms we present a general framework for ADAEs in chapter 2. Furthermore we discuss guiding questions for the treatment of ADAEs and familiarize the reader briefly with relevant index concepts and solvability approaches that have been developed for ADAEs in recent literature. In chapter 3 we concentrate on finite dimensional DAEs first and derive global solvability and perturbation results for a specific class of semi-linear

DAEs fulfilling monotonicity properties on certain subspaces. In this context also the MNA equations are investigated. Chapter 4 is devoted to an abstract approach for non-linear ADAEs involving a strongly monotone operator. It is an extension of the linear case presented in [Tis04]. Unique solvability is shown by means of convergent Galerkin solutions. Additionally we analyze the system with regard to perturbations on the right hand side. In chapter 5 two prototype coupled systems are presented and investigated with regard to solvability, perturbations and convergence of Galerkin solutions. The obtained results are applied to exemplary coupled systems in circuit simulation. Finally we summarize our results and give an outlook for future research concerning the study of ADAEs. The appendix covers the essential functional analytic material which is used throughout this thesis.

2. A general setting for ADAEs

Numerous mathematical models in science and engineering give rise to systems comprising partial differential equations (PDEs), ordinary differential equations (ODEs), algebraic equations and differential-algebraic equations (DAEs). They occur frequently in application areas such as electric circuit simulation ([Tis04]), flexible multibody systems ([Sim98, Sim00]) or chemical engineering ([CM96b]). Depending on space variables and time, these systems are usually discretized in space first by the method of lines and then result in a finite dimensional DAE. So it is common to use the term partial differential-algebraic equation (PDAE) for these kind of systems. In literature often the term PDAE is used when essentially the semi-discretized system is studied. As our focus will be on the analysis of the original infinite dimensional system we prefer the term abstract differential-algebraic equation (ADAE) as introduced in [Tis04, Rei06]. Further common terms are degenerate differential equations ([FY99]) or descriptor systems ([Rei06]).

This chapter will provide a brief overview of ADAEs. After having assembled some fundamental mathematical notation concerning function spaces we present a general framework for ADAEs. We formulate guiding questions for studying ADAEs with respect to solvability, perturbation analyis and convergence of solutions of the discretized system to the solution of the original ADAE. As a classification tool we recapitulate the ADAE Index and the Perturbation Index. Finally, we present the relevant approaches in literature for treating general ADAEs and coupled systems in circuit simulation.

2.1. Function spaces and norms

In this section we introduce some mathematical notation concerning relevant function spaces to be used throughout this thesis. For keeping this introductory part as brief as possible we recommend [Ada75, Zei86, Wer05, Eva08] for more functional analytic background material.

A real Banach space V is a complete normed vector space (on $\mathbb{R}$). The corresponding norm in V is denoted by $\|\cdot\|_V$. The dual space of V, denoted by V^*, is the set of all linear and continuous functionals on V, i.e.

$$V^* := \{f : V \to \mathbb{R} | \ f \text{ is linear and bounded}\}.$$

Furthermore we write

$$\langle f, v \rangle_V := f(v) \quad \forall v \in V$$

and with the norm

$$\|f\|_{V^*} := \sup_{\|v\|_V \leq 1} |\langle f, v \rangle|$$

the dual space V^* is a real Banach space. Let $(v_n) \subseteq V$ be a sequence, then we say that v_n converges to $v \in V$ (short: $v_n \to v$) as $n \to \infty$, if it converges in the norm, i.e. if $\|v_n - v\|_V \to 0$ as $n \to \infty$. The sequence (v_n) converges weakly to v (short: $v_n \rightharpoonup v$) as $n \to \infty$ if

$$\langle f, v_n \rangle_V \to \langle f, v \rangle_V \quad \forall f \in V^* \quad \text{as } n \to \infty.$$

The real Banach space V is a Hilbert space if there is a scalar product on V which we denote by $(\cdot | \cdot)_V$. If $V = \mathbb{R}^n$, $n \in \mathbb{N}$, the Euclidean scalar product and its induced norm are denoted by

$$(x \,|\, y) := x^\top y, \quad \|x\| := \sqrt{x^\top x} \qquad \forall x, y \in \mathbb{R}^n.$$

A matrix $A \in \mathbb{R}^{m \times n}$ can be measured by the induced operator norm

$$\|A\|_* := \sup_{x \in \mathbb{R}^n, x \neq 0} \frac{\|Ax\|}{\|x\|}.$$

Let V, W be Banach spaces and $U \subseteq V$. Then we write $C(U, W)$ for the set of all continuous functions $f : U \to W$. Usually continuity is defined on open sets. Therefore, if U is arbitrary, we have to be specific. In this case we mean that $f \in C(U, W)$, if f can be extended locally to a continuous function, i.e. for every $v \in U$ there exists an open neighborhood $U(v) \subseteq V$ such that f can be extended to a continuous function on $U(v)$, cf. [Zei86, Definition 4.22]. Similarly, we denote by $C^1(U, W)$ the set of all continuously differentiable functions on U. If $f : V \to W$ is linear and continuous we write $f \in L(V, W)$ and $f \in L(V)$ if $V = W$. Note that in the case that f is linear, f is continuous if and only if f is bounded. We set

$$\|f\|_{L(V,W)} := \sup_{v \in V, v \neq 0} \frac{\|f(v)\|_W}{\|v\|_V}.$$

Another important notion is Lipschitz continuity. Let $f : V \times M \to W$ be a map and $M \subseteq X$ be a subset of the Banach space X. Then f is Lipschitz continuous on V if there is $L > 0$ such that

$$\|f(v, z) - f(\bar{v}, z)\|_W \leq L \|v - \bar{v}\|_V \quad \forall v, \bar{v} \in V, z \in M.$$

Note here that L does not depend on $z \in M$.

For a set $\Omega \subseteq \mathbb{R}^n$ we denote by $\overline{\Omega}$ the closure of Ω, its interior by $\overset{\circ}{\Omega}$ and the boundary by $\partial \Omega$. If Ω is open we denote by $C(\Omega)$ the set of continuous functions $f : \Omega \to \mathbb{R}$, by $C^k(\Omega)$, $k \in \mathbb{N}$, the set of k-times continuously differentiable functions on Ω and by $C^\infty(\Omega)$ the set of all infinitely often continuously differentiable functions on Ω. We write $f \in C(\overline{\Omega})$, if $f \in C(\Omega)$ can be extended continuously to the boundary. Of importance is also the space of test functions $C_0^\infty(\Omega)$ which consists of all $f \in C^\infty(\Omega)$ having compact support. We will also encounter these spaces for $\Omega = \mathcal{I} \subseteq \mathbb{R}$ being a compact time interval. Then $C(\mathcal{I}, W)$ is a Banach space with the norm

$$\|f\|_{C(\mathcal{I}, W)} = \max_{t \in \mathcal{I}} \|f(t)\|_W \quad \forall f \in C(\mathcal{I}, W)$$

and if $W = \mathbb{R}^n$ we write $\|f\|_{C(\mathcal{I}, W)} = \|f\|_\infty$. We also encounter the space $C^1(\mathcal{I}, W)$ and if $f \in C^1(\mathcal{I}, W)$ we denote the (time) derivative by $f'(t)$ or $\frac{\mathrm{d}}{\mathrm{d}t} f(t)$. Especially the latter will be used for functions depending on other variables as well. For a deeper treatment of continuous and continuously differentiable functions we refer to [Zei86].

Next, we address spaces of integrable functions. We set for $1 \leq p < \infty$

$$L_p(\Omega) = \left\{ v : \Omega \to \mathbb{R} \text{ measurable} \mid \|v\|_{L_p(\Omega)} < \infty \right\}, \quad \|v\|_{L_p(\Omega)} := \left(\int_\Omega |v(x)|^p \, \mathrm{d}x \right)^{\frac{1}{p}}$$

where the integral is to be understood in the Lebesgue sense. $L_p(\Omega)$ is a Banach space and for $p = 2$ it becomes a Hilbert space with the standard scalar product

$$(u \mid v)_{L_2(\Omega)} = \int_\Omega u(x) v(x) \mathrm{d}x.$$

Note that elements of $L_p(\Omega)$ are strictly speaking not functions but equivalence classes, see e.g. [Wer05]. Two functions $u, v \in L_p(\Omega)$ are equivalent if they differ only on a subset of Ω of (Lebesgue) measure zero. Therefore a pointwise evaluation of these functions is not well-defined, but as long as we work with integral expressions we can treat them as functions. We also state the space of essentially bounded functions on Ω:

$$L_\infty(\Omega) = \left\{ v : \Omega \to \mathbb{R} \text{ measurable} \mid \operatorname*{ess\,sup}_{x \in \Omega} |v(x)| < \infty \right\}$$

The definition of Lebesgue spaces is a starting point for defining many important classes of functions spaces. Especially Sobolev spaces play a fundamental role in the treatment of PDEs, see [Ada75, Jos07]. For defining weak derivatives of real valued functions the space of locally integrable functions $L_{1,\mathrm{loc}}(\Omega)$ is essential. A function $v : \Omega \to \mathbb{R}$ is locally integrable if and only if $v \in L_1(K)$ for every compact subset $K \subseteq \Omega$. We say that $u \in L_{1,\mathrm{loc}}(\Omega)$ has a weak derivative in direction x_i with $x = (x_1, \ldots, x_n)^\top$ if there is $w_i \in L_{1,\mathrm{loc}}(\Omega)$ such that

$$\int_\Omega u \partial_i v \, \mathrm{d}x = - \int_\Omega w_i v \, \mathrm{d}x \quad \forall v \in C_0^\infty(\Omega). \tag{2.1}$$

We have omitted the argument x here and $\partial_i v$ denotes the classical derivative of v in the direction of x_i. We use the same symbol for the weak derivative and write $w_i = \partial_i u$ and

$$\nabla u = (\partial_1 u \ldots \partial_n u).$$

It is well-known that, if $u \in C^1(\Omega)$, the weak derivatives exist and coincide with the classical ones. Then equation (2.1) is simply the integration by parts formula. Note that weak derivatives can also be defined for higher order, but this is of no importance in this thesis. The space

$$H^1(\Omega) = \{u \in L_2(\Omega) | \ \partial_i u \in L_2(\Omega), \ 1 \le i \le n\}$$

is a Hilbert space with

$$\|u\|_{H^1(\Omega)} = \left(\|u\|_{L_2(\Omega)}^2 + \sum_{i=1}^n \|\partial_i u\|_{L_2(\Omega)}^2 \right)^{\frac{1}{2}},$$

$$(u|v)_{H^1(\Omega)} = (u|v)_{L_2(\Omega)} + \sum_{i=1}^n (\partial_i u | \partial_i v)_{L_2(\Omega)}.$$

We also use the notation:

$$\int_\Omega \nabla u \nabla v \, \mathrm{d}x = \sum_{i=1}^n (\partial_i u | \partial_i v)_{L_2(\Omega)}$$

Another important space is

$$H_0^1(\Omega) = \{u \in H^1(\Omega) | \ u|_{\partial\Omega} = 0\}.$$

Note here that the condition $u|_{\partial\Omega} = 0$ has to be understood in the sense of the trace operator, see [Ada75]. We also have that $H_0^1(\Omega)$ is the completion of $C_0^\infty(\Omega)$ with respect to (w.r.t.) the norm $\|\cdot\|_{H^1(\Omega)}$. $H_0^1(\Omega)$ is a Hilbert space and we write $(H_0^1(\Omega))^* = H^{-1}(\Omega)$.

So far we have only considered Lebesgue spaces for real valued functions. They, however, can also be generalized to Lebesgue spaces with values in Banach spaces. An introduction is given in Appendix A.3, where also generalized derivatives and evolution triples are discussed. Additionally in the appendix we cover fundamentals with regard to projectors on Banach spaces, the Theorem of Carathéodory, the Theorem of Browder-Minty and the Theorem of Arzelà-Ascoli.

2.2. A general form for ADAEs

A general form of nonlinear ADAEs was proposed in [Tis04] based on first observations made in [LMT01]. Let spaces V, W, Z, Z_w be real Banach or Hilbert spaces. Consider

$$\mathcal{A}\frac{\mathrm{d}}{\mathrm{d}t}\mathcal{D}(u(t),t) + \mathcal{B}(u(t),t) = 0, \quad t \in \mathcal{I} \tag{2.2}$$

on a fixed time interval $\mathcal{I} := [t_0, T] \subseteq \mathbb{R}$. Equation (2.2) is to be understood as an operator equation with

$$\mathcal{A}: Z_w \to W, \quad \mathcal{D}(\cdot,t): V \to Z, \quad \mathcal{B}(\cdot,t): V \to W, \quad t \in \mathcal{I}.$$

Note that in general $\mathcal{A}$ may depend on u and t as well and that we also allow the operators $\mathcal{A}$, $\mathcal{D}$ and $\mathcal{B}$ to be defined on subsets of V and Z_w respectively. In this thesis we only consider the case of $\mathcal{A}$ being constant. Note also that $\mathcal{A}$ and $\mathcal{D}$ may be singular. The time derivative can be a classical or a generalized derivative depending on the underlying spaces.

The form (2.2) is especially intended for studying the structure of coupled systems. These systems consist of DAEs and PDEs which exchange information via certain variables and coupling terms. Coupled systems in circuit simulation will be discussed at the end of this chapter. Many PDEs and DAEs themselves fit naturally in the framework of (2.2). In the following we present two examples, one being a standard second order PDE and the other one being a simple DAE describing the behavior of a simple electric circuit. As coupled systems comprise both PDEs and DAEs this is a good starting point. For examples of coupled sytems we refer to [LMT13, chapter 12] and to chapter 5 of this thesis where we study prototypes of coupled systems.

Example 2.1 (Standard second order PDE).
The following is taken from [Eva08, Jos07]. Let Ω be an open and bounded subset of $\mathbb{R}^n$ and $\mathcal{I} := [t_0, T] \subseteq \mathbb{R}$ be a fixed interval. We set $\Omega_T := \Omega \times (t_0, T]$ and study the problem

$$u' + \mathcal{L}u = f \quad \text{in } \Omega_T, \tag{2.3a}$$
$$u = 0 \quad \text{on } \partial\Omega \times \mathcal{I}, \tag{2.3b}$$
$$u = u_0 \quad \text{on } \Omega \times \{t_0\} \tag{2.3c}$$

where $f: \Omega_T \to \mathbb{R}$, $u_0 : \Omega \to \mathbb{R}$ are given and $u : \overline{\Omega}_T \to \mathbb{R}$ is the unknown ($u = u(x,t)$). $\mathcal{L}$ denotes for every t a second order partial differential operator in divergence form

$$\mathcal{L}u = -\sum_{i,j=1}^{n} \partial_j(a_{ij}(x,t)\partial_i u) + \sum_{i=1}^{n} b_i(x,t)\partial_i u + c(x,t)u$$

with given coefficient functions a_{ij}, b_i and c $(i, j = 1, \ldots, n)$. Note that for $a_{ii} = 1$, $a_{ij} = 0$ $(i \neq j)$, $b_i = c = f = 0$ we obtain the heat equation, because $\mathcal{L}$ reduces to the Laplace operator, i.e. $\mathcal{L} = -\Delta$. We assume that $a_{ij} = a_{ji}, b_i, c \in L_\infty(\Omega_T)$ for all i, j, $f \in L_2(\Omega_T)$ and $u_0 \in L_2(\Omega)$. The bilinear form

$$b(u, v, t) := \int_\Omega \sum_{i,j=1}^n a_{ij}(x, t) \partial_i u \partial_j v + \sum_{i=1}^n b_i(x, t) \partial_i v + c(x, t) u v \, dx$$

is defined for $u, v \in H_0^1(\Omega)$ and for almost all (f.a.a.) $t \in \mathcal{I}$. We write

$$u(t)(x) = u(x, t) \quad \text{and} \quad f(t)(x) = f(x, t).$$

Thus we identify u, f with $u : \mathcal{I} \to H_0^1(\Omega)$ and $f : \mathcal{I} \to L_2(\Omega)$. Following the standard procedure of multiplying by a test function, integrating over Ω and finally applying integration by parts, we obtain the weak formulation of (2.3):

$$\langle u'(t), v \rangle_{H_0^1(\Omega)} + b(u(t), v, t) = (f(t)|v)_{L_2(\Omega)} \quad \forall v \in H_0^1(\Omega), \text{ f.a.a. } t \in \mathcal{I} \tag{2.4a}$$

$$u(t_0) = u_0 \in L_2(\Omega) \tag{2.4b}$$

where u' is to be understood as a generalized derivative. Note that

$$H_0^1(\Omega) \subseteq L_2(\Omega) \subseteq H^{-1}(\Omega)$$

forms an evolution triple. If

$$u \in W_2^1(\mathcal{I}; H_0^1(\Omega), L_2(\Omega)) := \left\{ w \in L_2(\mathcal{I}, H_0^1(\Omega)) \middle| \; w' \in L_2(\mathcal{I}, H^{-1}(\Omega)) \right\}$$

fulfills (2.4) we say that u is a weak solution. Because the embedding

$$W_2^1(\mathcal{I}; H_0^1(\Omega), L_2(\Omega)) \subseteq C(\mathcal{I}, L_2(\Omega))$$

is continuous, the initial value condition (2.4b) makes sense. For relevant background material on evolution triples, generalized derivatives and the involved Lebesgue spaces we refer to Appendix A.3 and [Zei90a, Eva08, Jos07]. Setting

$$V = Z = H_0^1(\Omega), \quad W = Z_w = V^* = H^{-1}(\Omega)$$

and

$$\mathcal{A} := \text{id}_{V^*}, \quad \mathcal{D} := \text{id}_V, \quad \langle \mathcal{B}(u, t), v \rangle_V := b(u, v, t) - (f(t)|v)_{L_2(\Omega)}$$

puts (2.4) into the form (2.2). id_V and id_{V^*} are the identity maps on V and V^* respectively.

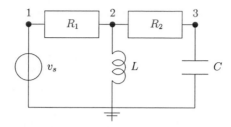

Figure 2.1.: Sample circuit with two resistors R_1 and R_2, a capacitor C, an inductor L and a voltage source v_s.

Example 2.2 (DAEs – sample circuit).
The form (2.2) is a generalization of so-called semi-linear or quasi-linear DAEs

$$A\frac{\mathrm{d}}{\mathrm{d}t}d(x,t) + b(x,t) = 0 \quad \text{on } \mathcal{D} \times \mathcal{I} \tag{2.5}$$

with $\mathcal{D} \subseteq \mathbb{R}^n$ being open and connected, $A \in \mathbb{R}^{n \times m}$, $d : \mathcal{D} \times \mathcal{I} \to \mathbb{R}^m$ and $b : \mathcal{D} \times \mathcal{I} \to \mathbb{R}^n$, cf. [LMT13]. Here $V = W = \mathbb{R}^n$ and $Z = Z_w = \mathbb{R}^m$. Electrical circuits can be described by a DAE and a small circuit is given in Figure 2.1. The corresponding equations read

$$g_{R,1}(e_1 - e_2, t) - j_V = 0$$
$$-g_{R,1}(e_1 - e_2, t) + g_{R,2}(e_2 - e_3, t) + j_L = 0$$
$$\frac{\mathrm{d}}{\mathrm{d}t}q_C(e_3, t) - g_{R,2}(e_2 - e_3, t) = 0$$
$$\frac{\mathrm{d}}{\mathrm{d}t}\phi_L(j_L, t) - e_2 = 0$$
$$-e_1 - v_s(t) = 0$$

with functions $g_{R,1}, g_{R,2}, q_C, \phi_L : \mathbb{R} \times \mathcal{I} \to \mathbb{R}$ describing the behavior of the resistors R_1, R_2, the capacitor C and the inductor L. The voltage source is described by the function $v_s : \mathcal{I} \to \mathbb{R}$. The variables are node potentials e_1, e_2, e_3, the current through the inductor j_L and the current through the voltage source j_V. The equations rely on the Modified Nodal Analysis (MNA) which will be described in chapter 3. With

$$x = (x_1, x_2, x_3, x_4, x_5)^\top = (e_1, e_2, e_3, j_L, j_V)^\top$$

we set

$$A := \begin{pmatrix} 0 & 0 \\ 0 & 0 \\ 1 & 0 \\ 0 & 1 \\ 0 & 0 \end{pmatrix}, \quad d(x,t) := \begin{pmatrix} q_C(x_3,t) \\ \phi_L(x_4,t) \end{pmatrix},$$

$$b(x,t) := \begin{pmatrix} g_{R,1}(x_1 - x_2, t) - x_5 \\ -g_{R,1}(x_1 - x_2, t) + g_{R,2}(x_2 - x_3, t) + x_4 \\ -g_{R,2}(x_2 - x_3, t) \\ -x_2 \\ -x_1 - v_s(t) \end{pmatrix}$$

to obtain the form (2.5).

A useful tool for classifying DAEs is its index as it gives an insight into the solution and perturbation behavior. There are several index concepts available for DAEs. To mention just a few we name the Kronecker Index ([GP83, GM86, LMT13]), the Differentiation Index ([Cam87, BCP96]), the Tractability Index ([GM86, Ria08, LMT13]), the Strangeness Index ([KM94, KM06]) and the Perturbation Index ([HLR89, HNW02]). For an overview of these and also other important index concepts for DAEs we refer to [Voi06, Meh12]. Regarding solvability they have proven fruitful for analyzing the structure of the DAE and thus identifying the inherent dynamics and all (hidden) constraints. So for higher index DAEs not only integration but also differentiation problems occur. Furthermore initial values cannot be chosen freely as it is the case for ODEs. They have to satisfy all constraints of the system, for an in-depth treatment we refer to [Est00, Bau12]. Furthermore DAEs of higher index generally represent ill-posed problems, i.e. small perturbations on the right hand side may result in large differences in the solution. This dependence is measured by the Perturbation Index and is very important for the numerical treatment of DAEs, cf. [LMT13].

A similar tool would be desirable for ADAEs which are defined on infinite dimensional spaces. A comprehensive theory for existence and uniqueness of solutions of (nonlinear) ADAEs does not exist. Even in the case of standard PDEs, where classifications in terms of elliptic, parabolic or hyperbolic PDEs exist, numerous approaches are used. For systems of PDEs or coupled systems the situation becomes even more complex. Therefore we propose the following guiding questions which arise naturally when studying ADAEs.

1. What conditions must be met for an ADAE of the form (2.2) to be solvable? When is the solution unique?

2. How does the solution change if the system is perturbed? Are there index criteria – similar to DAEs – which describe the perturbation behavior?

3. How should an ADAE of the form (2.2) be discretized? Is the solution of the discretized system unique and does it converge to the exact solution?

The last question is concerned with the numerical treatment of ADAEs. Being usually defined on infinite dimensional function spaces they have to be discretized in space and in time. The standard approach is to discretize in space first and then in time which is well-known under the term method of lines. After space discretization for many mixed systems (e.g. in chemical engineering) a DAE is obtained, cf. [CM96a, CM96b]. The function spaces can, for example, be approximated with the Galerkin approach. The Galerkin equations yield a DAE and it is important to achieve convergence of the Galerkin solutions to the solution of the original ADAE.

In the following we will concentrate on ADAEs where the operators $\mathcal{A}$ and $\mathcal{D}$ are potentially singular. For the treatment of DAEs we refer to the standard textbooks [BCP96, KM06, Ria08, LMT13] and the references therein. PDEs are extensively discussed in literature, see e.g. [Zei90a, Zei90b, Jos07, Eva08]. Some index concepts for DAEs mentioned previously have been generalized to the setting of ADAEs. We will briefly discuss these concepts with regard to the questions stated above.

ADAE Index

For DAEs the Tractability Index has proven fruitful for investigating the structure of possibly nonlinear DAEs in terms of solvability and perturbation estimates, cf. [LMT13]. The so-called ADAE Index was first proposed in [LMT01]. We follow here the more general version presented in [Tis04, Bod07] which is also covered in [LMT13]. The ADAE Index naturally extends the Tractability Index for DAEs of the form (2.5) to operator equations of the form (2.2) and is also based on linearizations. More precisely, we consider here the operator equation (2.2), where the operators $\mathcal{A}$, $\mathcal{D}(\cdot, t)$ and $\mathcal{B}(\cdot, t)$ act on the real Hilbert spaces $V = H$, $Z = Z_w$ and W for all $t \in \mathcal{I}$. We assume the Fréchet derivatives of the operators $\mathcal{B}(\cdot, t)$ and $\mathcal{D}(\cdot, t)$ to exist, i.e. there exist linear continuous operators $\mathcal{B}_0(u, t)$ and $\mathcal{D}_0(u, t)$ such that

$$\mathcal{B}(u + h, t) - \mathcal{B}(u, t) - \mathcal{B}_0(u, t)h = o(\|h\|_V) \quad \text{as } h \to 0$$
$$\mathcal{D}(u + h, t) - \mathcal{D}(u, t) - \mathcal{D}_0(u, t)h = o(\|h\|_V) \quad \text{as } h \to 0$$

for all h in some neighborhood of zero and all $u \in V$, $t \in \mathcal{I}$. We say that

$$r(h) = o(\|h\|_V) \text{ as } h \to 0 \quad \Leftrightarrow \quad \frac{r(h)}{\|h\|_V} \to 0 \text{ as } h \to 0.$$

We assume that the spaces $\operatorname{im} \mathcal{D}_0(u, t)$ and $\ker \mathcal{D}_0(u, t)$ do not depend on u and t. We set $N_0 := \ker \mathcal{D}_0(u, t)$ which is a closed subspace in V because $\mathcal{D}_0(u, t)$ is bounded. Furthermore we assume $\mathcal{A}$ and $\mathcal{D}_0$ to be well-matched, i.e. the transversality condition

$$\ker \mathcal{A} \oplus \operatorname{im} \mathcal{D}_0(u, t) = Z$$

holds for all $u \in V$ and $t \in \mathcal{I}$. With $\oplus$ we denote the topological sum and so $\operatorname{im} \mathcal{D}_0(u,t)$ is closed. As a consequence there is a constant projection operator $\mathcal{R} : Z \to Z$ satisfying

$$\operatorname{im} \mathcal{R} = \operatorname{im} \mathcal{D}_0(u,t), \quad \ker \mathcal{R} = \ker \mathcal{A}$$

for all $u \in V$ and $t \in \mathcal{I}$. For DAEs this is known under the term of a properly stated leading term, cf. [Mär03, HM04], [LMT13, chapter 3]. We set

$$\mathcal{G}_0(u,t) := \mathcal{A}\mathcal{D}_0(u,t) \quad \forall u \in V, \ t \in \mathcal{I}.$$

and it holds

$$\operatorname{im} \mathcal{G}_0(u,t) = \operatorname{im} \mathcal{A}\mathcal{R} = \operatorname{im} \mathcal{A} \quad \text{and} \quad \ker \mathcal{G}_0(u,t) = \ker \mathcal{R}\mathcal{D}_0(u,t) = \ker \mathcal{D}_0(u,t).$$

The natural solution space for the ADAE (2.2) is

$$C_{\mathcal{D}}^1(\mathcal{I}, V) := \left\{ u \in C(\mathcal{I}, V) \middle| \mathcal{D}(u(\cdot), \cdot) \in C^1(\mathcal{I}, Z) \right\}$$

but unfortunately it is not linear. As pointed out in [Tis04, Remark 4.2] for a linearization

$$\mathcal{A}\frac{\mathrm{d}}{\mathrm{d}t}\mathcal{D}_0(u_*(t), t) + \mathcal{B}_0(u_*(t), t) = 0$$

of (2.2) at $u_* \in C_{\mathcal{D}}^1(\mathcal{I}, V)$ the natural solution space reads

$$C_{\mathcal{D}_0}^1(\mathcal{I}, V) := \left\{ u \in C(\mathcal{I}, V) \middle| \mathcal{D}_0(u_*(\cdot), \cdot)u(\cdot) \in C^1(\mathcal{I}, Z) \right\}.$$

Under additional smoothness assumptions the spaces $C_{\mathcal{D}}^1(\mathcal{I}, V)$ and $C_{\mathcal{D}_0}^1(\mathcal{I}, V)$ coincide, cf. [LMT13, Proposition 12.1]. We are now able to define the ADAE Index.

Definition 2.3 (ADAE Index).
Let an ADAE of the form (2.2) with all assumptions made above be given. We say that the ADAE has

- ADAE Index 0, if $\dim N_0 = 0$ and $\overline{\operatorname{im} \mathcal{G}_0} = W$;

- ADAE Index 1, if $\dim N_0 > 0$ and there is a projection operator $\mathcal{Q}_0 : V \to V$ onto the constant space $\ker \mathcal{G}_0(u,t)$ such that the operator

  $$\mathcal{G}_1(u,t) := \mathcal{G}_0(u,t) + \mathcal{B}_0(u,t)\mathcal{Q}_0$$

  is injective and $\overline{\operatorname{im} \mathcal{G}_1(u,t)} = W$;

- ADAE Index 2, if $\dim N_0 > 0$, $\dim(\ker \mathcal{G}_1(u,t)) > 0$ and there are projection operators $\mathcal{Q}_0 : V \to V$ onto $\ker \mathcal{G}_0(u,t)$ and $\mathcal{Q}_1(u,t) : V \to V$ onto $\ker G_1(u,t)$ and the operator

  $$\mathcal{G}_2(u,t) := \mathcal{G}_1(u,t) + \mathcal{B}_0(u,t)\mathcal{P}_0\mathcal{Q}_1(u,t)$$

  is injective and $\overline{\operatorname{im} \mathcal{G}_2(u,t)} = W$. $\mathcal{P}_0$ is the complementary projector of $\mathcal{Q}_0$.

It can be shown that the ADAE Index 1 definition does not depend on the choice of the projector $\mathcal{Q}_0$. We also stress that the ADAE Index requires the spaces Z and Z_w to be equal. This, in general, does not need to be the case, if e.g. the weak formulation of a PDE as in Example 2.1 is considered. Furthermore the weak derivative of the solution is only square integrable instead of being continuous. In chapter 4 we follow an abstract solvability approach for ADAEs and encounter a solution which is only square integrable in time. In the particular case we will give a sensible definition of the operators $\mathcal{G}_0$ and $\mathcal{G}_1$ and prove that they satisfy the corresponding conditions. Furthermore the ADAE Index definition coincides with the one for the Tractability Index (cf. [LMT13]) if the underlying spaces are $V = \mathbb{R}^n$, $Z = Z_w = \mathbb{R}^k$ and $W = \mathbb{R}^m$.

The ADAE Index (in the infinite dimensional setting) was examined in several test cases in [LMT01] and [LMT13, chapter 12]. Further index results were shown for coupled systems in circuit simulation with distributed semiconductor devices, cf. [Sch02, Tis04, Bod07]. There the known index results for the semi-discretized system were validated for the original infinite dimensional system. As for DAEs an index definition should give us information about the perturbation behavior. A connection between the ADAE Index and a perturbation result is not known so far as it is for certain classes of nonlinear DAEs having Tractability Index 1 or 2, cf. e.g. [Bau12, Theorems 2.49 and 2.50]. We now define the Perturbation Index for ADAEs.

Perturbation Index for ADAEs

The Perturbation Index for DAEs, cf. [HLR89], has also been extended to ADAEs. We present here the definition given in [Bod07]. A similar definition can be found in [CM96b] and in [RA05] for linear ADAEs. Let $u_* \in C^1_{\mathcal{D}_0}(\mathcal{I}, V)$ be the unique solution to (2.2) with initial value $u_*(t_0) = u_0$. Additionally, the perturbed initial value problem

$$\mathcal{A}\frac{\mathrm{d}}{\mathrm{d}t}\mathcal{D}(u(t), t) + \mathcal{B}(u(t), t) = \delta(t), \quad t \in \mathcal{I}, \tag{2.6a}$$

$$u(t_0) = u_0^\delta \tag{2.6b}$$

is considered where the perturbation $\delta \in C^{k-1}(\mathcal{I}, W)$, its derivatives $\delta^{(i)}$ for $1 \leq i < k$ and $u_0^\delta - u_0 =: \delta_0 \in V$ are sufficiently small. Let u_δ be the unique solution to (2.6) and k be the smallest number for which an estimate of the form

$$\max_{t \in \mathcal{I}} \|u_\delta(t) - u_*(t)\|_V \leq c \left(\|\delta_0\|_V + \sum_{i=0}^{k-1} \max_{t \in \mathcal{I}} \|\delta^{(i)}(t)\|_W \right)$$

with a constant $c > 0$ holds. Then the ADAE (2.2) has (continuous) Perturbation Index k. We wrote $\delta^{(0)} = \delta$.

For higher index DAEs it is well-known that even small perturbations on the right hand side and in the initial conditions can lead to a large difference in the solution. This is the case for DAEs with index higher than 1 because then the derivative of δ occurs on the right hand side of the estimate. Furthermore it is assumed that the solution u_* is in $C_D^1(\mathcal{I}, V)$. As pointed out already for the ADAE Index this cannot be always expected especially when considering weak formulations of PDE systems. Therefore a more general definition considering also weak solutions in time is proposed.

Definition 2.4 (Perturbation Index).
Let $u_* \in S \subseteq X$ be the unique solution to (2.2) in some solution space S with given initial value $u_*(t_0) = u_0 \in H$. Let X, Y, H be Banach spaces where no time derivatives are measured. Additionally consider the perturbed initial value problem

$$\mathcal{A}\frac{\mathrm{d}}{\mathrm{d}t}\mathcal{D}(u(t), t) + \mathcal{B}(u(t), t) = \delta(t), \quad t \in \mathcal{I}, \tag{2.7a}$$

$$u(t_0) = u_0^\delta \tag{2.7b}$$

and assume that the perturbation $\delta \in Y$, its derivatives $\delta^{(i)} \in Y$ for $1 \le i < k$ and $u_0^\delta - u_0 =: \delta_0 \in H$ are sufficiently small. Let $u_\delta \in S$ be the unique solution to (2.7) and k be the smallest number for which an estimate of the form

$$\|u_\delta - u_*\|_X \le c\left(\|\delta_0\|_H + \sum_{i=0}^{k-1}\left\|\delta^{(i)}\right\|_Y\right) \tag{2.8}$$

with a constant $c > 0$ holds. Then we say that (2.2) has Perturbation Index k.

Definition 2.4 comprises the definition of the Perturbation Index for continuous functions because setting $S = C_{D_0}^1(\mathcal{I}, V)$, $X = Y = C(\mathcal{I}, V)$ and $H = V$ results in the situation above. Less regular spaces are also possible, e.g. in the case of Example 2.1. Here we would have $S = W_2^1(\mathcal{I}; H_0^1(\Omega), L_2(\Omega))$, $X = L_2(\mathcal{I}, H_0^1(\Omega))$, $Y = L_2(\mathcal{I}, H^{-1}(\Omega))$ and $H = L_2(\Omega)$. We will encounter a similar situation in chapters 4 and 5.

ADAEs in literature

We assemble here approaches for ADAEs concerning generalizations of other index concepts, solvability results and also theoretical results for coupled sytems in circuit simulation. Most of the approaches have in common that they rely on continuous or continuously differentiable solutions to (2.2). In [MB00] a generalization of the Differentiation Index is presented which also takes spatial derivatives into account. Furthermore in [LSEL99] a differential spatial and a differential time index for linear time independent ADAEs is formulated. The approach is based on the Laplace transform and Fourier transform. A generalization of the Perturbation Index for PDAEs is defined on a quite formal level in [CM96b]. Starting out from a classical formulation of a linear constant PDE system with initial and boundary conditions various examples are examined. It is pointed out that perturbation analysis should not only take the right hand side and initial conditions but also boundary conditions and the spatial domain into account. Note that for the form (2.7) perturbations δ on the right hand side may depend on the space variable as well because $\delta(t) \in W$.

In [RA05] ADAEs of the form

$$\mathcal{A}u'(t) + \mathcal{B}u(t) = r(t), \quad t \in \mathcal{I}, \tag{2.9}$$

with linear time-invariant operators $\mathcal{A}$, $\mathcal{B}$ and a sufficiently smooth function r are considered. In contrast to the earlier approaches mentioned in the previous paragraph here a weak formulation (in space) of the underlying PDE system is investigated. However, with regard to the time t the solution is assumed to be continuously differentiable. Perturbation estimates are obtained because of the requirement that the operator $\mathcal{B}$ satisfies a certain Gårding inequality. Additionally, an alternative Perturbation Index definition for linear ADAEs is given in [RA05, Definition 3.3] where also perturbations of the boundary conditions and derivatives of perturbations of the initial value are considered.

A theoretical approach for analyzing linear time-independent ADAEs of the form (2.9) was developed by Reis, cf. [RT05, Rei06, Rei07]. Here $V = Z = Z_w$ and W are Hilbert spaces and $\mathcal{A} : V \to W$ is a bounded linear operator wheras $\mathcal{B} : \text{dom}(\mathcal{B}) \subseteq V \to W$ is a closed linear operator. With $\text{dom}(\mathcal{B})$ we mean the domain of the operator $\mathcal{B}$. Motivated by the Kronecker normal form for finite dimensional DAEs with constant coefficients a decoupled version of (2.9) is achieved:

$$\begin{pmatrix} N & 0 \\ 0 & I \\ 0 & 0 \end{pmatrix} \frac{\mathrm{d}}{\mathrm{d}t} \begin{pmatrix} u_1(t) \\ u_2(t) \end{pmatrix} + \begin{pmatrix} I & K \\ 0 & L \\ 0 & M \end{pmatrix} \begin{pmatrix} u_1(t) \\ u_2(t) \end{pmatrix} = \begin{pmatrix} r_1(t) \\ r_2(t) \\ r_3(t) \end{pmatrix} \tag{2.10}$$

Here N is linear, bounded and nilpotent, i.e. there is $\nu > 0$ such that $N^\nu = 0$, $N^{\nu-1} \neq 0$. The number ν is called the ADAE Index and can be compared to the Kronecker Index for finite dimensional DAEs. The proof of the decoupled form is based on projectors $\mathcal{Q}_i$ similar to the ones needed for the definition in the ADAE Index above. This gives some justification for the ADAE Index for ADAEs of the form (2.9), cf. [RT05, Theorem 4.1]. The main differences between (2.10) and the finite dimensional case are the boundary control operator M and the coupling operator K. Having the decoupling (2.10), the set of consistent initial values can be parameterized and perturbation results can be obtained, cf. [Rei06, Rei07]. The appearance of the boundary and coupling operator leads to additional difficulties for the parametrization of consistent initial values. An initial value not only has to fulfill (hidden) algebraic constraints but also some additional (hidden) boundary constraints.

Considering the theory of evolution equations there are two major approaches for solving evolution equations, namely the semigroup and the Galerkin approach, see e.g. [Lio69, GGZ74, Zei90a, Zei90b, Sho97, Emm04, Rou05]. In [Tis04] the Galerkin approach for linear ADAEs with a strongly monotone, bounded and time-dependent operator $\mathcal{B}$ has been proven. Furthermore the system has been analyzed with regard to perturbations on the right hand side and the ADAE Index. In chapter 4 we study this approach again but allow the operator $\mathcal{B}$ to be nonlinear.

Although we focus on the Galerkin approach throughout this treatise, we give some remarks on the treatment of ADAEs with the semigroup approach. Fundamental work is collected in [FY99] where ADAEs are also called degenerate differential equations. Linear systems of the form

$$\mathcal{A}\frac{\mathrm{d}}{\mathrm{d}t}\mathcal{D}u(t) + \mathcal{B}u(t) = \mathcal{A}r(t), \quad t \in \mathcal{I}, \tag{2.11a}$$

$$\mathcal{D}u(t_0) = x_0 \in \operatorname{im}\mathcal{D} \tag{2.11b}$$

are considered where $\mathcal{A}$, $\mathcal{D}$ and $\mathcal{B}$ are linear operators and $V = Z = Z_w = W$ is a Hilbert space. If $\mathcal{A} = \operatorname{id}_V$ the operators $\mathcal{D}$ and $\mathcal{B}$ are defined on linear subspaces $\operatorname{dom}(\mathcal{B}) \subseteq \operatorname{dom}(\mathcal{D}) \subseteq V$. If $r \in C^1(\mathcal{I}, V)$, a certain passivity condition holds (with a constant $\beta \in \mathbb{R}$) and the operator

$$\lambda_0 \mathcal{D} + \mathcal{B} : \operatorname{dom}(\mathcal{B}) \subseteq V \to V \text{ is bijective for some } \lambda_0 > \beta \tag{2.12}$$

then the system (2.11) has a unique (classical) solution $u : \mathcal{I} \to \operatorname{dom}(\mathcal{B})$ such that $\mathcal{D}u \in C^1(\mathcal{I}, V)$ and $\mathcal{B}u \in C(\mathcal{I}, V)$, cf. Theorem 2.9, [FY99]. In the proof equation (2.11) is transformed with $\mathcal{M} := -\mathcal{B}\mathcal{D}^{-1}$ into a multivalued differential equation of the form

$$x'(t) \in \mathcal{M}x(t) + f(t), \quad t \in \mathcal{I},$$
$$x(t_0) = x_0$$

for a suitable f. We remark that the operator $\mathcal{D}^{-1}$ is defined as a multivalued function satisfying

$$\mathcal{D}^{-1}u = \{v \in \operatorname{im}(\operatorname{dom}(\mathcal{D}))|\, \mathcal{D}v = u\} \quad \forall u \in \operatorname{im}\mathcal{D}.$$

Similar solvability results were obtained in the cases that $\mathcal{A} = \mathcal{D}^*$ or $\mathcal{D} = \operatorname{id}_V$. The approach relies mainly on the passivity condition and on (2.12) and is meant for dealing with hyperbolic problems. We remark that the linear operators $\mathcal{A}, \mathcal{D}, \mathcal{B}$ may be unbounded. In [Tis04] it is pointed out that these ADAEs – considering the finite dimensional case – have Tractability Index 1 and have to satisfy a certain constraint which is implied by the passivity condition.

In addition to that the semigroup approach has also been extended to linear time dependent systems of the form

$$\mathcal{A}(t)\frac{\mathrm{d}}{\mathrm{d}t}(\mathcal{D}(t)u(t)) + \mathcal{B}(t)u(t) = \mathcal{A}(t)r(t), \quad t \in \mathcal{I},$$

$$\mathcal{D}(t_0)u(t_0) = x_0 \in \operatorname{im}\mathcal{D}$$

with $\mathcal{A}(t) = \operatorname{id}_V$ or $\mathcal{D}(t) = \operatorname{id}_V$, cf. [FY99]. Unique solvability relies on certain uniform bounds of terms involving the operator

$$\mathcal{R}(\lambda, t) := (\lambda\mathcal{D}(t) + \mathcal{B}(t))^{-1}$$

for all $t \in \mathcal{I}$ and $\lambda \in \Sigma \subseteq \mathbb{C}$ with Σ being a specific region in the complex plane. In practice these conditions are usually quite difficult to verify. From the theory of finite dimensional DAEs it is also known that time dependent matrix pencils are not as well-suited for solving DAEs as in the constant case. For example in [BCP96, chapter 2.4] it is mentioned that in general neither regular matrix pencils $(\mathcal{D}(t), \mathcal{B}(t))$ imply solvability nor solvability implies a matrix pencil to be regular. This is also stated in [GM86] and illustrating examples can be found in [BCP96, Examples 2.4.1 and 2.4.1] and [Tis04]. For further approaches to linear degenerate evolution equations we refer to [FY99] and the references therein. An overview of spectral methods for studying degenerate differential operator equations can be found in [Rut07].

Nonlinear approaches for solving ADAEs of the form

$$\frac{\mathrm{d}}{\mathrm{d}t}(\mathcal{D}u(t)) + \mathcal{B}u(t) = f(t, u(t)), \quad t \in \mathcal{I}$$

are studied in [FP86, FR99, RV03, RK04, Rut08]. Here $\mathcal{D}$ and $\mathcal{B}$ are closed linear operators from a Banach space V into a Banach space W and f is a nonlinear term satisfying certain smoothness properties. The solvability results are local except in [RK04] where global solvability is obtained requiring strong smoothness assumptions for f. The main condition for all these approaches is that the resolvent $(\mathcal{D} + \mu\mathcal{B})^{-1}$ has a pole in $\mu = 0$ of maximal order 2. This resolvent condition, however, is difficult to verify in applications because the existence of the resolvent and the pole property has to be validated.

Apart from the results presented above some solvability results also exist for special (nonlinear) ADAEs, e.g. multiphysically modeled systems in circuit simulation. Here the circuit is modeled by the equations of the MNA which we will present in chapter 3. The resulting system is a DAE where the variables are the node potentials and certain currents of the electrical network. To simulate semiconductors in high frequencies correctly a distributed modeling is essential. A common model is given by the drift-diffusion equations (DD-equations), cf. [Moc83, Sel84, Gaj85, GG86, Mar86, Gaj93]. The DD-equations are a system of one elliptic and two parabolic PDEs for every semiconductor. The coupling between the DAE and the PDE system is two-directional. On the one hand evaluations of boundary integrals of the contacts of the semiconductor devices have an influence on the DAE and on the other hand boundary conditions are described by functions depending on the node potentials, cf. [Tis04, MT11]. Solvability results of these coupled systems are given in [ABGT03, ABG05, ABG10] for certain topological assumptions, which guarantee the network equations to have Tractability Index 1. They are based on applications of the Schauder or Banach Fixpoint Theorem and well-known solvability results for the DD-equations, cf. [Mar86]. This system has been studied with regard to perturbations in [Bod07]. In [Gün01] a linear hyperbolic system simulating transmission lines is coupled to the MNA equations. The solvability of the coupled system is proven by the Galerkin approach and perturbation estimates are derived as well. It is pointed out that the index of the Galerkin equations coincides with the one of the coupled system. Using other discretization schemes the semi-discretization of the ADAE may act like a deregularization (increasing the index) or regularization (decreasing the index), cf. [Gün00]. In [Bar04] a model is presented which couples the network equations to the one dimensional heat equation in order to simulate heating effects in a circuit. Here solvability is also investigated and the proof mainly relies on fixpoint arguments. In [Cul09] the higher dimensional case is considered along with the Galerkin approach. We will discuss this system in chapter 5.

2.3. Conclusion

In this chapter basic notation concerning relevant function spaces has been collected. Then we presented a general form (2.2) for abstract differential-algebraic equations (ADAEs) from [LMT01, LMT13] which is based on operators being defined on certain Banach or Hilbert spaces. Furthermore we proposed guiding questions for treating ADAEs concerning solvability, perturbation analysis and convergence of solutions of discretization methods. The ADAE Index (Definition 2.3) from [LMT01, Tis04] and the Perturbation Index for ADAEs from [Bod07] are presented. The definition of the Perturbation Index was slightly generalized in order to apply it also to solutions which are not necessarily continuous or continuously differentiable in time, cf. Definition 2.4.

Additionally, we covered relevant results from the literature investigating index concepts, solvability and perturbation behavior of ADAEs. These approaches mainly focus on linear time-invariant operators, see [CM96b, LSEL99, FY99, RA05, Rei06], except for some nonlinear solvability results which are based on a semigroup approach, see e.g. [FR99, RK04]. The Galerkin approach and the convergence of the Galerkin solutions to the solution of the original system for a certain class of linear time dependent ADAEs are discussed in [Tis04]. This approach will be extended to the nonlinear case in chapter 4.

Having applications from circuit simulation in mind we are especially interested in coupled systems. We mentioned main results concerning the solvability and perturbation behavior of coupled systems in circuit simulation which add semiconductors [Tis04, ABGT03, ABG10, Bod07] or transmission lines ([Gün01]) to the circuit or simulate additional heating effects ([Bar04, Cul09]). These systems are very complex and for the derivation of the results the underlying structure of the systems has to be exploited. There is no general approach available for treating (nonlinear) coupled systems with regard to solvability, perturbation behavior and convergence of Galerkin solutions. Therefore it is essential to develop prototype coupled systems which cover already a certain coupling structure. This task will be pursued in chapter 5.

3. Solvability of nonlinear DAEs with monotonicity properties

In the last three decades differential-algebraic equations (DAEs) have become a major mathematical tool for modeling and simulation in many application areas including electrical circuit simulation, mechanical multibody systems, gas and water network simulation and many others, see [KM06, Ria08, LMT13]. From a theoretical point of view solvability results and sensitivity with regard to perturbations for DAEs are of great interest. Solvability results for nonlinear DAEs are in general local, i.e. given in a neighborhood of a given initial value, cf. [KM06, LMT13]. The well-known Implicit Function Theorem in combination with a suitable transformation, reduction or decoupling of the original system is a key to obtain these results. A differential-geometric approach can be found in [Rei91]. Some global results exist which ensure the existence of a solution on a given fixed interval. But they require strong smoothness assumptions and uniform bounds of certain inverse matrices, see [GM86, RK04, CC07]. These conditions are quite difficult to verify in actual applications. In the context of coupled systems which may be approximated by Galerkin equations, for example, global solvability results for DAEs are desirable because otherwise the existence interval of solutions may vary with the Galerkin step n.

In circuit simulation a standard tool for describing a circuit's behavior are the equations of the Modified Nodal Analysis (MNA). These equations represent a nonlinear DAE and local solvability results and perturbation estimates are known which are based on the concept of the Tractability Index, cf. [Tis99, ET00, Tis04]. It is also well-known that the Tractability Index of the MNA equations depends on the topology of the electrical network. Furthermore, there it is also assumed that the constitutive relations of the basic circuit elements satisfy certain monotonicity (or passivity) conditions. Here the question arises whether these monotonicity conditions can be used directly to prove a global solvability result and a perturbation estimate. This will be investigated in this chapter.

The results of this chapter were developed in cooperation with Jansen and Tischendorf and can also be found in [JMT12]. In section 3.1 we present important definitions and results for solving nonlinear algebraic equations, both globally and locally, based on the concept of strong monotonicity. The dependence of the solution on other components will be important in this context. In section 3.2 we will use these results to obtain unique solvability results for a certain class of semi-linear DAEs. It will be shown that these DAEs have Perturbation Index 1. In section 3.3 we describe briefly the derivation of the MNA equations. Finally, we apply the solvability and perturbation results of section 3.2 to the MNA equations in the topological index 1 case.

3.1. Preliminaries

In this section we assemble more or less well-known tools for solving nonlinear algebraic equations with (local) monotonicity properties. We will use these tools extensively in the later sections of this chapter.

Definition 3.1 (Lipschitz continuity).
Let a function $f : \mathbb{R}^n \times \mathbb{R}^m \to \mathbb{R}^k$, $n, m, k \in \mathbb{N}$, be given. Then

(i) f is Lipschitz continuous w.r.t. the first argument x if there is a constant $L_f > 0$ such that

$$\|f(x_2, y) - f(x_1, y)\| \leq L_f \|x_2 - x_1\|, \quad \forall x_1, x_2 \in \mathbb{R}^n, y \in \mathbb{R}^m.$$

(ii) f is locally Lipschitz continuous w.r.t. the first argument x in $(x_0, y_0) \in \mathbb{R}^n \times \mathbb{R}^m$ if there are neighborhoods $U(x_0) \subseteq \mathbb{R}^n, V(y_0) \subseteq \mathbb{R}^m$ and a scalar $L > 0$ such that

$$\|f(x_2, y) - f(x_1, y)\| \leq L \|x_2 - x_1\|, \quad \forall x_1, x_2 \in U(x_0), y \in V(y_0).$$

(iii) f is locally Lipschitz continuous on $\mathbb{R}^n \times \mathbb{R}^m$ if f is locally Lipschitz continuous for all $(x, y) \in \mathbb{R}^n \times \mathbb{R}^m$.

We indicate that Definition 3.1 (i) can also be found under the term global Lipschitz continuity when stressing that it holds for all $x_1, x_2 \in \mathbb{R}^n$ or under the term uniform Lipschitz continuity when stressing that L_f is independent of y. However we will use the term Lipschitz continuity as defined in 3.1 (i). If f is continuously differentiable w.r.t. x then f is Lipschitz continuous w.r.t. x if and only if there is an $L > 0$ such that $\|f_x(x, y)\|_* \leq L$ for all $(x, y) \in \mathbb{R}^n \times \mathbb{R}^m$.

Definition 3.2 (Strong monotonicity).
Let a function $f : \mathbb{R}^n \times \mathbb{R}^m \to \mathbb{R}^m$ be given. Then

(i) f is strongly monotone w.r.t. the second argument y if there is a scalar $\mu_f > 0$ such that

$$(f(x, y_2) - f(x, y_1) | y_2 - y_1) \geq \mu_f \|y_2 - y_1\|^2, \quad \forall x \in \mathbb{R}^n, y_1, y_2 \in \mathbb{R}^m.$$

(ii) f is locally strongly monotone w.r.t. the second argument y in $(x_0, y_0) \in \mathbb{R}^n \times \mathbb{R}^m$ if there are neighborhoods $U(x_0) \subseteq \mathbb{R}^n$, $V(y_0) \subseteq \mathbb{R}^m$ and a scalar $\mu > 0$ such that

$$(f(x, y_2) - f(x, y_1) | y_2 - y_1) \geq \mu \|y_2 - y_1\|^2, \quad \forall x \in U(x_0), y_1, y_2 \in V(y_0).$$

(iii) f is locally strongly monotone on $\mathbb{R}^n \times \mathbb{R}^m$ if f is locally strongly monotone for all $(x, y) \in \mathbb{R}^n \times \mathbb{R}^m$.

For strongly monotone functions the following Corollary is an immediate consequence.

Corollary 3.3.
Let a function $f : \mathbb{R}^n \times \mathbb{R}^m \to \mathbb{R}^m$ be given.

(i) If f is strongly monotone w.r.t. y, then

$$\|f(x, y_2) - f(x, y_1)\| \geq \mu_f \|y_2 - y_1\|, \quad \forall x \in \mathbb{R}^n, y_1, y_2 \in \mathbb{R}^m.$$

(ii) If f is locally strongly monotone w.r.t. y in (x_0, y_0), then

$$\|f(x, y_2) - f(x, y_1)\| \geq \mu \|y_2 - y_1\|, \quad \forall x \in U(x_0), y_1, y_2 \in V(y_0).$$

Here $U(x_0)$, $V(y_0)$ are the same neighborhoods and μ, μ_f are the same constants as in Definition 3.2.

PROOF:
For all $(x, y_1), (x, y_2) \in \mathbb{R}^n \times \mathbb{R}^m$ and $(x, y_1), (x, y_2) \in U(x_0) \times V(y_0)$, respectively, it holds

$$\widetilde{\mu} \|y_2 - y_1\|^2 \leq (f(x, y_2) - f(x, y_1) | y_2 - y_1) \leq \|f(x, y_2) - f(x, y_1)\| \|y_2 - y_1\|$$

by the Cauchy-Schwarz inequality with $\widetilde{\mu} = \mu_f$ or $\widetilde{\mu} = \mu$. □

Therefore strong monotonicity can be regarded as a counter part of Lipschitz continuity. The Jacobian of a strongly monotone function is bounded from below as we will see in the next lemma.

Lemma 3.4.
Let the continuous function $f : \mathbb{R}^n \times \mathbb{R}^m \to \mathbb{R}^m$ be continuously differentiable w.r.t. y and let $f_y(x, y)$ be the Jacobian of f w.r.t. y at the point (x, y). Then it holds:

(i) f is strongly monotone w.r.t. y if and only if the map $z \mapsto f_y(x, y)z$ is strongly monotone w.r.t. z, i.e. there is a $\mu > 0$ such that

$$(z | f_y(x, y)z) \geq \mu \|z\|^2, \quad \forall (x, y) \in \mathbb{R}^n \times \mathbb{R}^m \ \forall z \in \mathbb{R}^m.$$

(ii) In the case of (i) $f_y(x,y)$ is bounded from below by μ, i.e.

$$\|f_y(x,y)\|_* \geq \mu, \quad \forall (x,y) \in \mathbb{R}^n \times \mathbb{R}^m.$$

PROOF:
(i) The proof can also be found in [OR70, p. 142]. But we state it here for completeness.
Let $x \in \mathbb{R}^n$ be arbitrary.
($\Rightarrow$) Let $z, y_1 \in \mathbb{R}^m, z \neq 0$, $s > 0$ and $y_2 := y_1 + sz$. With the mean value theorem we obtain

$$
\begin{aligned}
s^2 \mu \|z\|^2 = \mu \|y_2 - y_1\|^2 &\leq (y_2 - y_1 | f(x,y_2) - f(x,y_1)) \\
&= (sz | \int_0^1 f_y(x, y_1 + tsz)\mathrm{d}t \; sz) \\
&= s^2 (z | \int_0^1 f_y(x, y_1 + tsz)\mathrm{d}t \; z).
\end{aligned}
$$

Dividing both sides by s^2 leaves the left side independent of s whereas the right one still is. Taking the limit $s \to 0$ we obtain

$$\mu \|z\|^2 \leq (z | \int_0^1 f_y(x, y_1 + tsz)\mathrm{d}t \; z) \to (z | f_y(x, y_1)z).$$

($\Leftarrow$) Let $y_1, y_2 \in \mathbb{R}^m$ then we have applying the mean value theorem:

$$
\begin{aligned}
(y_1 - y_2 | f(x,y_1) - f(x,y_2)) &= (y_1 - y_2 | \int_0^1 f_y(x, y_2 + t(y_1 - y_2))\mathrm{d}t(y_1 - y_2)) \\
&= \int_0^1 (y_1 - y_2 | f_y(x, y_2 + t(y_1 - y_2))(y_1 - y_2))\mathrm{d}t \\
&\geq \mu \|y_1 - y_2\|^2
\end{aligned}
$$

(ii) With (i) and the Cauchy-Schwarz inequality we obtain for all $(x,y) \in \mathbb{R}^n \times \mathbb{R}^m$ that

$$\mu \|z\|^2 \leq (z | f_y(x,y)z) \leq \|z\| \, \|f_y(x,y)z\| \leq \|z\|^2 \, \|f_y(x,y)\|_*.$$

$\square$

Example 3.5.
The function

$$f : \mathbb{R} \to \mathbb{R}, \quad f(x) = c_1 x^3 + c_2 x, \quad c_1 \geq 0, \; c_2 > 0$$

is strongly monotone whereas the function

$$g : \mathbb{R} \to \mathbb{R}, \quad g(x) = e^x$$

is not. This can be seen with Lemma 3.4.

A function $f : \mathbb{R}^n \to \mathbb{R}^n$, which is Lipschitz continuous and strongly monotone, is bijective. This is a consequence of the well-known Theorem of Zarantonello in the setting of real Hilbert spaces. It is a generalization of Cea's Lemma for bilinear, coercive and bounded forms, cf. [Emm04, Satz 3.5.2] or [Zei90b, Theorem 25.B]. From the strong monotonicity of f it follows easily that f^{-1} is Lipschitz continuous. However, dropping the Lipschitz continuity of f and having only continuity implies global solvability as well. This is formulated in [OR70, Theorem 6.4.4]. Again this is a special version of the well-known more general Browder-Minty Theorem for monotone operators on Banach spaces, cf. Theorem A.14, [Zei90b, Theorem 26.A] or [Rou05, chapter 2.3]. We state it here for the space $\mathbb{R}^n$ and will refer to it in the later chapters of this thesis. For a proof cf. [OR70, Theorem 6.4.4].

Theorem 3.6.
Let $f : \mathbb{R}^n \to \mathbb{R}^n$ be continuous and strongly monotone. Then the equation

$$f(x) = y$$

has a unique solution $x \in \mathbb{R}^n$ for each $y \in \mathbb{R}^n$.
Furthermore the inverse function $f^{-1} : \mathbb{R}^n \to \mathbb{R}^n$ is Lipschitz continuous.

Lemma 3.7.
Let $\mathcal{I} \subseteq \mathbb{R}$ be an interval and $f : \mathbb{R}^n \times \mathbb{R}^m \times \mathcal{I} \to \mathbb{R}^m$ be a continuous function. Then for all $(x,t) \in \mathbb{R}^n \times \mathcal{I}$ the equation

$$f(x, y, t) = 0 \tag{3.1}$$

has a unique solution $y \in \mathbb{R}^m$ if f is strongly monotone w.r.t. y and Lipschitz continuous w.r.t. x. The solution depends on (x,t) and we write $y = \psi(x,t)$ with the continuous function $\psi : \mathbb{R}^n \times \mathcal{I} \to \mathbb{R}^m$ which is Lipschitz continuous w.r.t. x.

PROOF:
The unique solvability is derived from Theorem 3.6 for all but fixed $(x,t) \in \mathbb{R}^n \times \mathcal{I}$ with the solution $y_{x,t}$. By setting $\psi(x,t) := y_{x,t}$ we obtain the solution function ψ. The continuity of ψ as well as the Lipschitz continuity w.r.t. x have to be checked.
Continuity. Let $(x_n, t_n) \in \mathbb{R}^n \times \mathcal{I}$ be a sequence with $(x_n, t_n) \to (x,t) \in \mathbb{R}^n \times \mathcal{I}$ as $n \to \infty$ and hence

$$f(x_n, \psi(x_n, t_n), t_n) = 0 = f(x, \psi(x,t), t).$$

We obtain with the strong monotonicity (scalar $\mu_f > 0$)

$$
\begin{aligned}
&\|\psi(x,t) - \psi(x_n, t_n)\| \\
&\leq \frac{1}{\mu_f} \|f(x_n, \psi(x,t), t_n) - f(x_n, \psi(x_n, t_n), t_n)\| \\
&= \frac{1}{\mu_f} \|f(x_n, \psi(x,t), t_n) - f(x, \psi(x,t), t)\| \to 0 \quad \text{as } n \to \infty
\end{aligned}
$$

because f is continuous. So ψ is continuous.

Lipschitz continuity. Let $x_1, x_2 \in \mathbb{R}^n$, $t \in \mathcal{I}$ and hence

$$f(x_1, \psi(x_1, t), t) = 0 = f(x_2, \psi(x_2, t), t).$$

Similar as before is follows with strong monotonicity

$$
\begin{aligned}
\|\psi(x_2, t) - \psi(x_1, t)\| &\leq \frac{1}{\mu_f} \|f(x_2, \psi(x_2, t), t) - f(x_2, \psi(x_1, t), t)\| \\
&= \frac{1}{\mu_f} \|f(x_1, \psi(x_1, t), t) - f(x_2, \psi(x_1, t), t)\| \\
&\leq \frac{L_f}{\mu_f} \|x_2 - x_1\|.
\end{aligned}
$$

The last line was obtained using the Lipschitz continuity of f ($L_f > 0$). $\qquad\square$

Remark 3.8 (Solution function).

A function ψ as in Lemma 3.7 will be called solution function. This means generally that there is a unique function ψ satisfying

$$y = \psi(x, t) \Leftrightarrow f(x, y, t) = 0.$$

A sufficient condition for having a continuous ψ which is Lipschitz continuous in x was given in Lemma 3.7. However another sufficient condition is based on the Theorem of Hadamard, cf. [OR70, Theorem 5.3.10]. It states that, if $g : \mathbb{R}^m \to \mathbb{R}^m$ is continuously differentiable and $\|g_y(y)^{-1}\|_* \leq \gamma < \infty$ for a $\gamma > 0$ for all $y \in \mathbb{R}^m$, then g is a homeomorphism of $\mathbb{R}^m$ onto $\mathbb{R}^m$. Consider (3.1) with f being continuously differentiable, $\|f_y(x, y, t)^{-1}\|_*$ being uniformly bounded and f being Lipschitz continuous w.r.t. x. Then for given (x, t) we obtain a solution $y = \psi(x, t)$. From the standard Implicit Function Theorem, cf. [OR70, Theorem 5.2.4] or [Zei86, Theorem 4.B], it follows now that ψ is continuous and

$$\psi_x(x, t) = -f_y(x, \psi(x, t), t)^{-1} f_x(x, \psi(x, t), t).$$

So $\|\psi_x(x, t)\|$ is uniformly bounded because f_y^{-1} and f_x are. The latter is bounded because f is Lipschitz continuous w.r.t. x. So ψ itself is Lipschitz continuous w.r.t. x.

The Implicit Function Theorem is a standard tool for solving nonlinear algebraic equations of the form

$$f(x, y) = 0$$

locally w.r.t. x around a point (x_0, y_0) with $f(x_0, y_0) = 0$. For a general overview we refer to [KP03]. In its standard version the continuous differentiability of f is required. Nevertheless there is an Implicit Function Theorem by Kumagai based on an observation of Jittorntrum, cf. [Jit78, Kum80], which does not require differentiability of f. More precisely we state the following.

Theorem 3.9 (cf. [Jit78, Kum80]).
Let $f : \mathbb{R}^n \times \mathbb{R}^m \to \mathbb{R}^m$ be continuous and $f(x_0, y_0) = 0$ for $x_0 \in \mathbb{R}^n$, $y_0 \in \mathbb{R}^m$. Suppose there exist neighborhoods $A \subseteq \mathbb{R}^n$ and $B \subseteq \mathbb{R}^m$ of x_0 and y_0, respectively, such that $f(x, \cdot) : B \to \mathbb{R}^m$ is locally one-to-one in y for all $x \in A$. Then there exist neighborhoods $A_0 \subseteq \mathbb{R}^n$ and $B_0 \subseteq \mathbb{R}^m$ of x_0 and y_0, respectively, such that for all $x \in A_0$ the equation

$$f(x, y) = 0$$

has a unique solution

$$y = \psi(x) \in B_0$$

where $\psi : A_0 \to B_0$ is continuous.

Here a function $f : B \subseteq \mathbb{R}^n \to \mathbb{R}^m$ is said to be locally one-to-one if for every point $y \in B$ there is a neighborhood $V \subseteq B$ of y such that $f|_V$ is injective. The following Lemma is an application of Theorem 3.9. It gives a local solution result in analogy to Lemma 3.7 using local strong monotonicity.

Lemma 3.10.
Let $f : \mathbb{R}^n \times \mathbb{R}^m \times \mathbb{R} \to \mathbb{R}^m$ be continuous and let there be $(x_0, y_0, t_0) \in \mathbb{R}^n \times \mathbb{R}^m \times \mathbb{R}$ with

$$f(x_0, y_0, t_0) = 0.$$

Furthermore let f be locally strongly monotone w.r.t. y and locally Lipschitz continuous w.r.t. x in (x_0, y_0, t_0).
Then there exist neighborhoods $A_0 \subseteq \mathbb{R}^n \times \mathbb{R}$, $B_0 \subseteq \mathbb{R}^m$ of (x_0, t_0) and y_0, respectively, such that

$$f(x, y, t) = 0$$

has a unique solution $y = \psi(x, t)$ where $\psi : A_0 \to B_0$ is locally Lipschitz continuous w.r.t. x in (x_0, t_0).

PROOF:
Applying Corollary 3.3 there exist neighborhoods $A \subseteq \mathbb{R}^n \times \mathbb{R}$, $B \subseteq \mathbb{R}^m$ of (x_0, t_0) and y_0, respectively, such that there are $\mu, L > 0$ with

$$\|f(x, y_2, t) - f(x, y_1, t)\| \geq \mu \|y_2 - y_1\|, \quad \forall (x, t) \in A, \ y_1, y_2 \in B, \tag{3.2}$$

$$\|f(x_2, y, t) - f(x_1, y, t)\| \leq L \|x_2 - x_1\|, \quad \forall (x_1, t), \ (x_2, t) \in A, \ y \in B. \tag{3.3}$$

Let $(x, t) \in A$ and $y_1, y_2 \in B$ with $f(x, y_2, t) = f(x, y_1, t)$, then

$$0 = \|f(x, y_2, t) - f(x, y_1, t)\| \geq \mu \|y_2 - y_1\|$$

because of (3.2) and hence $y_1 = y_2$. So f is locally one-to-one in y for all $(x,t) \in A$. Because of Theorem 3.9 there exist neighborhoods $A_0 \subseteq A$, $B_0 \subseteq B$ of (x_0, t_0) and y_0, respectively, such that for all $(x,t) \in A_0$ the equation

$$f(x, y, t) = 0$$

has a unique solution

$$y = \psi(x,t) \in B_0$$

where $\psi : A_0 \to B_0$ is continuous. Let $(x_1, t), (x_2, t) \in A_0$ and hence

$$f(x_1, \psi(x_1, t), t) = 0 = f(x_2, \psi(x_2, t), t).$$

It follows

$$
\|\psi(x_2, t) - \psi(x_1, t)\| \overset{(3.2)}{\leq} \frac{1}{\mu} \|f(x_2, \psi(x_2, t), t) - f(x_2, \psi(x_1, t), t)\|
$$

$$
= \frac{1}{\mu} \|f(x_1, \psi(x_1, t), t) - f(x_2, \psi(x_1, t), t)\|
$$

$$
\overset{(3.3)}{\leq} \frac{L}{\mu} \|x_2 - x_1\|
$$

and so ψ is locally Lipschitz continuous. $\qquad\square$

In this section we have briefly summarized solvability results of algebraic equations which are (locally) strongly monotone and (locally) Lipschitz continuous. We extended these results to solvability results of parameter dependent algebraic equations in Lemmata 3.7 and 3.10. They are very important for solving certain DAEs with monotonicity properties in the next section.

3.2. Solvability results for DAEs

First, we specify the class of DAEs we are going to investigate. Let $\mathcal{I} \subseteq \mathbb{R}$ be a compact interval and $\mathcal{D} \subseteq \mathbb{R}^n$ be open and connected. Let the equation

$$A \frac{\mathrm{d}}{\mathrm{d}t} d(z, t) + b(z, t) = 0 \qquad (3.4)$$

with $A \in \mathbb{R}^{n \times m}$, $d \in C^1(\mathcal{D} \times \mathcal{I}, \mathbb{R}^m)$ and $b \in C(\mathcal{D} \times \mathcal{I}, \mathbb{R}^n)$ be given. We call (3.4) a semi-linear or quasi-linear DAE, cf. [LMT13]. Notice here that we already assume in the formulation that the matrix A is constant. Furthermore (3.4) is a special version of the general ADAE (2.2). For notational reasons we write z instead of $z(t)$.

Definition 3.11 (Properly stated leading term, cf. [LMT13]).
A DAE of the form (3.4) has a properly stated leading term on $\mathcal{D} \times \mathcal{I}$, if rk $(\text{im } d_z)$ is constant on $\mathcal{D} \times \mathcal{I}$, im d_z has a continuously differentiable basis in $\mathbb{R}^m$ and the transversality condition

$$\ker A \oplus \text{im } d_z(z,t) = \mathbb{R}^m, \quad \forall (z,t) \in \mathcal{D} \times \mathcal{I}$$

holds.

Furthermore we set

$$\mathcal{M}_0(t) := \{z \in \mathcal{D} | \; b(z,t) \in \text{im } A\}, \quad t \in \mathcal{I} \tag{3.5}$$

which is the obvious constraint set of (3.4). The flow of the DAE is restricted to $\mathcal{M}_0(t)$, cf. [Mär03, LMT13].

Assumption 3.12 (Basic assumptions for (3.4)).
Consider a DAE of the form (3.4). We assume that

(i) the DAE (3.4) has a properly stated leading term,

(ii) $\ker d_z(z,t)$ is independent of $(z,t) \in \mathcal{D} \times \mathcal{I}$.

Note that the second assumption implies that $(\ker d_z(z,t))^\perp$ is also independent of (z,t). In order to find a solution to (3.4) a decoupling step is useful to identify algebraic and differential solution components. The following two lemmata describe a way of decoupling DAEs relying on an orthonormal bases decomposition of certain subspaces. Given a function $f : \mathbb{R}^n \to \mathbb{R}^n$ it is well-known that for projectors $Q, \widetilde{Q} \in \mathbb{R}^{n \times n}$ (see Appendix A.1) and their complementary projectors P, $\widetilde{P}$, respectively, it holds:

$$f(x) = 0 \Leftrightarrow \widetilde{P}f(Px + Qx) = 0 \text{ and } \widetilde{Q}f(Px + Qx) = 0 \quad \forall x \in \mathbb{R}^n$$

The variables Px, Qx as well as the two equations are given in $\mathbb{R}^n$. This means the system size and the number of variables is doubled compared to the original system. However, the intrinsic dimension of Px, for example, is only $\dim(\text{im } P) \leq n$. With the following approach it is possible to formulate equations with variables which are equivalent to $f(x) = 0$ and are given in their intrinsic dimension. This idea was developed by Jansen, cf. [Jan13], and can be found in [JMT12] as well.

Lemma 3.13 (Projector and basis functions).
Let $M \in \mathbb{R}^{m \times n}$. Define

$$n_x := \dim((\ker M)^\perp), \quad n_y := \dim(\ker M)$$

and let

$$B_x := \{p_1, \dots, p_{n_x}\}, \quad B_y := \{q_1, \dots, q_{n_y}\}$$

be orthonormal bases of $(\ker M)^{\perp}$ *and of* $\ker M$, *respectively. Furthermore we set*

$$p := \left(p_1, \ldots, p_{n_x}\right) \in \mathbb{R}^{n \times n_x}, \quad q := \left(q_1, \ldots, q_{n_y}\right) \in \mathbb{R}^{n \times n_y}.$$

We have the following.

(i) $P := pp^{\top}$ *is an orthogonal projector with* $\operatorname{im} P = (\ker M)^{\perp}$,

(ii) $Q := qq^{\top}$ *is an orthogonal projector with* $\operatorname{im} Q = \ker M$,

(iii) P *is the complementary projector of* Q, *i.e.* $P = I - Q$,

(iv) *Let* $\mathcal{I} \subseteq \mathbb{R}$ *be a compact interval and* $\mathcal{D} \subseteq \mathbb{R}^n$ *be open and connected. Let be* $f \in C(\mathcal{D} \times \mathcal{I}, \mathbb{R}^m)$ *where the Jacobian* $f_z(z,t)$ *exists for all* $(z,t) \in \mathcal{D} \times \mathcal{I}$. *Assuming* $\ker f_z(z,t) = \ker M$ *to be independent of* (z,t) *and using the notation above it then follows that* $f(z,t) = f(Pz,t)$ *holds for all* $(z,t) \in \mathcal{D} \times \mathcal{I}$ *with* $sz + (1-s)Pz \in \mathcal{D}$, $s \in [0,1]$.

PROOF:
(i) P is a projector because

$$P^2 = pp^{\top}pp^{\top} = pI_n p^{\top} = P$$

and $P^{\top} = P$. Since $p^{\top} \in \mathbb{R}^{n_x \times n}$ has full row rank and $\operatorname{im} p = (\ker M)^{\perp}$ we get

$$\operatorname{im} P = \operatorname{im} pp^{\top} = \operatorname{im} p = (\ker M)^{\perp}.$$

(ii) $Q^2 = Q$, $Q^{\top} = Q$ and $\operatorname{im} Q = \ker M$ follow analogously.
(iii) The matrix $B := (p \quad q) \in \mathbb{R}^{n \times n}$ is orthonormal and therefore it holds $BB^{\top} = I_n$. This is equivalent to

$$\sum_{k=1}^{n} b_{ik}b_{jk} = \begin{cases} 1, & i = j \\ 0, & i \neq j \end{cases} \quad \forall 1 \leq i,j \leq n$$

with $B = (b_{ij})$. Then for all $1 \leq i,j \leq n$:

$$\sum_{k=1}^{n} b_{ik}b_{jk} = \sum_{k=1}^{n_x} b_{ik}b_{jk} + \sum_{k=n_x+1}^{n} b_{ik}b_{jk} = \sum_{k=1}^{n_x} (p_k)_i(p_k)_j + \sum_{k=1}^{n_y} (q_k)_i(q_k)_j.$$

It follows that $pp^{\top} + qq^{\top} = I_n$ and hence $P + Q = I_n$.
(iv) Since $\operatorname{im} Q = \ker M = \ker f_z(z,t)$ for all $(z,t) \in \mathcal{D} \times \mathcal{I}$ we see that

$$f_z(z,t)Q = 0 \text{ and } f_z(z,t)P = f_z(z,t).$$

Applying the mean value theorem $(sz + (1 - s)Pz \in \mathcal{D})$ to the difference gives

$$f(z, t) - f(Pz, t) = \int_0^1 f_z(Pz + s(1 - P)z, t) \mathrm{d}s (I - P)z$$

$$= \int_0^1 \underbrace{f_z(Pz + s(1 - P)z, t)Q}_{=0} \mathrm{d}s \, z = 0,$$

see here [Mär03]. □

We can now use Lemma 3.13 for decoupling a DAE of the form (3.4).

Lemma 3.14 (Decoupling).
Let $\mathcal{I} \subseteq \mathbb{R}$ be a compact interval and $\mathcal{D} \subseteq \mathbb{R}^n$ be open. Consider a DAE of the form (3.4) under the Assumption 3.12 with p, q *as in Lemma 3.13 for the matrix $d_z(z, t)$. Furthermore let $sz + (1 - s)\mathrm{pp}^\top z \in \mathcal{D}$, $s \in [0, 1]$ for all $z \in \mathcal{D}$. Define*

$$\mathcal{D}_x := \{x \in \mathbb{R}^{n_x} | \, \exists z \in \mathcal{D} : x = \mathrm{p}^\top z\},$$
$$\mathcal{D}_y := \{y \in \mathbb{R}^{n_y} | \, \exists z \in \mathcal{D} : y = \mathrm{q}^\top z\}.$$

Then (3.4) can be formulated equivalently as

$$\frac{\mathrm{d}}{\mathrm{d}t} m(x, t) = f(x, y, t) \tag{3.6a}$$

$$0 = g(x, y, t) \tag{3.6b}$$

with $f \in C(\mathcal{D}_x \times \mathcal{D}_y \times \mathcal{I}, \mathbb{R}^{n_x})$, $g \in C(\mathcal{D}_x \times \mathcal{D}_y \times \mathcal{I}, \mathbb{R}^{n_y})$, $m \in C^1(\mathcal{D}_x \times \mathcal{I}, \mathbb{R}^{n_x})$ and $m_x(x, t)$ being non-singular for all $(x, t) \in \mathcal{D}_x \times \mathcal{I}$. Here being formulated equivalently means that

$$(x_*, y_*) \in C^1(\mathcal{I}, \mathbb{R}^{n_x}) \times C(\mathcal{I}, \mathbb{R}^{n_y})$$

is a solution of (3.6) if and only if

$$z_* = \mathrm{p}x_* + \mathrm{q}y_* \in C(\mathcal{I}, \mathbb{R}^n)$$

is a solution of (3.4) with $\mathrm{p}^\top z_* \in C^1(\mathcal{I}, \mathbb{R}^{n_x})$.

PROOF:
It can be checked that $\mathcal{D}_x$ and $\mathcal{D}_y$ are open. Let v and w be as in Lemma 3.13 for the

matrix $A^\top$. With the help of Assumption 3.12 and Lemma 3.13 it follows

$$A\frac{\mathrm{d}}{\mathrm{d}t}d(z,t) + b(z,t) = 0$$

$$\Leftrightarrow \quad A\frac{\mathrm{d}}{\mathrm{d}t}d(Pz,t) + b((P+Q)z,t) = 0$$

$$\Leftrightarrow \quad A\frac{\mathrm{d}}{\mathrm{d}t}d(\mathrm{p}(\mathrm{p}^\top z),t) + b(\mathrm{p}(\mathrm{p}^\top z) + \mathrm{q}(\mathrm{q}^\top z),t) = 0$$

$$\Leftrightarrow \quad A\frac{\mathrm{d}}{\mathrm{d}t}d(\mathrm{p}x,t) + b(\mathrm{p}x + \mathrm{q}y,t) = 0$$

$$\overset{(*)}{\Leftrightarrow} \quad \begin{aligned} \mathrm{v}^\top A\frac{\mathrm{d}}{\mathrm{d}t}d(\mathrm{p}x,t) + \mathrm{v}^\top b(\mathrm{p}x+\mathrm{q}y,t) &= 0 \\ \mathrm{w}^\top b(\mathrm{p}x+\mathrm{q}y,t) &= 0 \end{aligned}$$

$$\Leftrightarrow \quad \begin{aligned} \frac{\mathrm{d}}{\mathrm{d}t}m(x,t) &= f(x,y,t) \\ 0 &= g(x,y,t) \end{aligned}$$

with

$$x := \mathrm{p}^\top z \in \mathbb{R}^{n_x}, \quad y := \mathrm{q}^\top z \in \mathbb{R}^{n_y}$$

and

$$\begin{aligned} m(x,t) &:= \mathrm{v}^\top A d(\mathrm{p}x,t), \\ f(x,y,t) &:= -\mathrm{v}^\top b(\mathrm{p}x + \mathrm{q}y,t), \\ g(x,y,t) &:= \mathrm{w}^\top b(\mathrm{p}x + \mathrm{q}y,t). \end{aligned}$$

Ad (*). ($\Rightarrow$) follows from left multiplication by $\mathrm{v}^\top$ and $\mathrm{w}^\top$ and observing that $\mathrm{w}^\top A = 0$ because

$$\ker \mathrm{w}^\top = (\operatorname{im} \mathrm{w})^\perp = \operatorname{im} \mathrm{v} = (\ker A^T)^\perp = \operatorname{im} A.$$

($\Leftarrow$) Multiplying the first line by v and the second by w gives

$$VA\frac{\mathrm{d}}{\mathrm{d}t}d(\mathrm{p}x,t) + Vb(\mathrm{p}x+\mathrm{q}y,t) = 0,$$

$$Wb(\mathrm{p}x+\mathrm{q}y,t) = 0$$

with the projectors $V := \mathrm{v}\mathrm{v}^\top$ and $W := \mathrm{w}\mathrm{w}^\top$. Adding these two equations and noticing that $VA = A$ gives the desired result. By the construction of x,y,m,f and g it is obvious that the DAEs (3.4) and (3.6) are equivalently formulated. □

We can now focus on the solvability of systems of the form (3.6) and subsequently apply the obtained results to DAEs of the form (3.4) via Lemma 3.14.

Theorem 3.15 (Global Solvability).
Let $\mathcal{I} \subseteq \mathbb{R}$ be a compact interval. Consider the system

$$\frac{\mathrm{d}}{\mathrm{d}t}m(x,t) = f(x,y,t) \tag{3.7a}$$

$$0 = g(x,y,t) \tag{3.7b}$$

with $f \in C(\mathbb{R}^{n_x} \times \mathbb{R}^{n_y} \times \mathcal{I}, \mathbb{R}^{n_x})$, $g \in C(\mathbb{R}^{n_x} \times \mathbb{R}^{n_y} \times \mathcal{I}, \mathbb{R}^{n_y})$, $m \in C^1(\mathbb{R}^{n_x} \times \mathcal{I}, \mathbb{R}^{n_x})$. If

(i) m is strongly monotone w.r.t. x,

(ii) f is Lipschitz continuous w.r.t. x and y,

(iii) g is Lipschitz continuous w.r.t. x and strongly monotone w.r.t. y,

then (3.7) has a unique solution $(x_*, y_*) \in C^1(\mathcal{I}, \mathbb{R}^{n_x}) \times C(\mathcal{I}, \mathbb{R}^{n_y})$ for every initial value $x_*(t_0) = x_0 \in \mathbb{R}^{n_x}$.

PROOF:
Existence. We can solve (3.7b) w.r.t. y due to Lemma 3.7 with a solution function $y = \psi(x,t)$ with $\psi \in C(\mathbb{R}^{n_x} \times \mathcal{I}, \mathbb{R}^{n_y})$ and ψ is Lipschitz continuous w.r.t. x. Inserting this expression into (3.7a) we obtain

$$\frac{\mathrm{d}}{\mathrm{d}t}m(x,t) = f(x, \psi(x,t), t) =: \widetilde{f}(x,t). \tag{3.8}$$

Clearly $\widetilde{f}$ is continuous and the Lipschitz continuity w.r.t. x follows from the Lipschitz continuity of ψ w.r.t. x and (ii).
Next we show an a priori estimate for any solution $x : J \to \mathbb{R}^{n_x}$ to (3.8) on an arbitrary subinterval $J := [t_0, T_J] \subseteq \mathcal{I}$. If x solves (3.8) we can integrate over $[t_0, t] \subseteq J$ and obtain

$$m(x(t), t) = m(x_0, t_0) + \int_{t_0}^{t} \widetilde{f}(x(s), s)\mathrm{d}s$$

with $x(t_0) = x_0$. Using the strong monotonicity of m we get

$$\begin{aligned}
&\mu \left\| x(t) - x_0 \right\| \\
\leq\ & \left\| m(x(t), t) - m(x_0, t) \right\| \\
\leq\ & \left\| m(x_0, t_0) - m(x_0, t) \right\| + \int_{t_0}^{t} \left\| \widetilde{f}(x(s), s) \right\| \mathrm{d}s \\
\leq\ & \left\| m(x_0, t_0) - m(x_0, t) \right\| + \int_{t_0}^{t} \left\| \widetilde{f}(x_0, s) \right\| \mathrm{d}s + \int_{t_0}^{t} \left\| \widetilde{f}(x(s), s) - \widetilde{f}(x_0, s) \right\| \mathrm{d}s \\
\leq\ & (T - t_0) \max_{\tau_f, \tau_m \in [t_0, T]} (\left\| m_t(x_0, \tau_m) \right\| + \|\widetilde{f}(x_0, \tau_f)\|) + L_{\widetilde{f}} \int_{t_0}^{t} \left\| x(s) - x_0 \right\| \mathrm{d}s
\end{aligned}$$

with $L_{\tilde{f}} > 0$. The last line is a consequence of the mean value theorem and the Lipschitz continuity of $\tilde{f}$. We conclude that there are constants $c_1, c_2 > 0$, independent of t, such that

$$\|x(t) - x_0\| \le c_1 + c_2 \int_{t_0}^t \|x(s) - x_0\|\, \mathrm{d}s.$$

Applying the Gronwall Lemma gives the desired a priori estimate

$$\|x(t) - x_0\| \le c_1 e^{c_2(T - t_0)} =: C$$

with $C > 0$ being independent of t. It is

$$\frac{\mathrm{d}}{\mathrm{d}t} m(x,t) = m_x(x,t)x' + m_t(x,t)$$

for x being continuously differentiable. Using Lemma 3.4 and the fact that m is continuously differentiable the map $z \mapsto m_x(x,t)z$ is continuous and strongly monotone w.r.t. z. So $m_x(x,t)$ is non-singular and the inverse $m_x^{-1}(x,t)$ is continuous because of Lemma 3.7.

Then (3.8) can be reformulated as

$$x' = m_x(x,t)^{-1}\left(\tilde{f}(x,t) - m_t(x,t)\right) =: \tilde{m}(x,t)$$

for $t \in \mathcal{I}$ with initial value $x(t_0) = x_0 \in \mathbb{R}^{n_x}$. The function $\tilde{m}$ is continuous as a combination of continuous functions. Hence we can apply the Peano Theorem, cf. [Zei86, Theorem 3.B]. We obtain a local solution $x \in C^1(J, \mathbb{R}^{n_x})$ on a subinterval $J \subseteq \mathcal{I}$ which can be extended to the whole interval $\mathcal{I}$ because of the a priori estimate above, cf. [Zei90b, p.801 (iii)]. So there is a solution $x_* \in C^1(\mathcal{I}, \mathbb{R}^{n_x})$ of (3.8) and by setting

$$y_*(\cdot) := \psi(x_*(\cdot), \cdot) \in C(\mathcal{I}, \mathbb{R}^{n_y})$$

we obtain a solution $(x_*, y_*) \in C^1(\mathcal{I}, \mathbb{R}^{n_x} \times \mathbb{R}^{n_y})$ to (3.7).

Uniqueness. For uniqueness let (x_1, y_1), (x_2, y_2) be two solutions which fulfill (3.7). So we have $x_i(t)$ fulfilling (3.8) and $y_i(t) = \psi(x_i(t), t)$ for $i = 1, 2, t \in \mathcal{I}$. Therefore we have on $\mathcal{I}$:

$$\frac{\mathrm{d}}{\mathrm{d}t} m(x_1(t), t) - \frac{\mathrm{d}}{\mathrm{d}t} m(x_2(t), t) = \tilde{f}(x_1(t), t) - \tilde{f}(x_2(t), t).$$

We have $x_1(t_0) = x_0 = x_2(t_0)$ and integration over $[t_0, t]$, $t \in \mathcal{I}$ yields

$$m(x_1(t), t) - m(x_2(t), t) = \int_{t_0}^t \tilde{f}(x_1(s), s) - \tilde{f}(x_2(s), s)\mathrm{d}s.$$

Using the strong monotonicity of m and the Lipschitz continuity of $\widetilde{f}$ we see that

$$
\begin{aligned}
\|x_1(t) - x_2(t)\| &\leq \frac{1}{\mu} \|m(x_1(t), t) - m(x_2(t), t)\| \\
&\leq \frac{1}{\mu} \int_{t_0}^{t} \left\| \widetilde{f}(x_1(s), s) - \widetilde{f}(x_2(s), s) \right\| ds \\
&\leq \frac{L_{\widetilde{f}}}{\mu} \int_{t_0}^{t} \|x_1(s) - x_2(s)\| ds.
\end{aligned}
$$

with $\mu > 0$. Gronwall's Lemma now reveals that $x_1(t) = x_2(t)$ for all $t \in \mathcal{I}$ and therefore $y_1(t) = \psi(x_1(t), t) = \psi(x_2(t), t) = y_2(t)$ for all $t \in \mathcal{I}$. □

An analogous result ensuring local unique solvability can be obtained as well.

Theorem 3.16 (Local solvability).
Let $\mathcal{I} \subseteq \mathbb{R}$ be a compact interval. Consider the system

$$
\frac{d}{dt} m(x, t) = f(x, y, t) \tag{3.9a}
$$

$$
0 = g(x, y, t) \tag{3.9b}
$$

with $f \in C(\mathbb{R}^{n_x} \times \mathbb{R}^{n_y} \times \mathcal{I}, \mathbb{R}^{n_x})$, $g \in C(\mathbb{R}^{n_x} \times \mathbb{R}^{n_y} \times \mathcal{I}, \mathbb{R}^{n_y})$ and $m \in C^1(\mathbb{R}^{n_x} \times \mathcal{I}, \mathbb{R}^{n_x})$ and

$$
(x_0, y_0) \in \mathcal{M}_0(t_0) = \left\{ (x, y) \in \mathbb{R}^{n_x} \times \mathbb{R}^{n_y} \mid g(x, y, t_0) = 0 \right\}.
$$

If

(i) m is locally strongly monotone w.r.t. x in (x_0, t_0),

(ii) f is locally Lipschitz continuous w.r.t. x and y in (x_0, y_0, t_0),

(iii) g is locally Lipschitz continuous w.r.t. x and locally strongly monotone w.r.t. y in (x_0, y_0, t_0),

then there exists $\tau > 0$ such that (3.9) has a unique solution (x, y) on $\mathcal{I}_* := [t_0, t_0 + \tau]$ with $(x, y) \in C^1(\mathcal{I}_*, \mathbb{R}^{n_x}) \times C(\mathcal{I}_*, \mathbb{R}^{n_y})$.

PROOF:
Existence. With the requirements (i)-(iii) we find a neighborhood

$$
U^0 = U_x(x_0) \times U_y(y_0) \times U_t(t_0)
$$

such that all of the requirements (i)-(iii) hold in U^0. In the following we use the shorter notation $U_x := U_x(x_0)$, $U_y := U_y(y_0)$ and $U_t := U_t(t_0)$. We can solve equation (3.9b) w.r.t. y due to Lemma 3.10 with a solution function $\psi \in C(U_x^0 \times U_t^0, U_y^0)$ with $y = \psi(x, t)$

and $U_x^0 \subseteq U_x, U_y^0 \subseteq U_y, U_t^0 \subseteq U_t$. ψ is locally Lipschitz continuous w.r.t. x in U_x^0. Inserting this expression into (3.9a) we obtain

$$\frac{\mathrm{d}}{\mathrm{d}t}m(x,t) = f(x, \psi(x,t), t) =: \widetilde{f}(x,t). \tag{3.10}$$

Clearly $\widetilde{f}$ is continuous and the local Lipschitz continuity w.r.t. x in U_x^0 follows from the local Lipschitz continuity of ψ w.r.t. x and (ii). It is

$$\frac{\mathrm{d}}{\mathrm{d}t}m(x,t) = m_x(x,t)x' + m_t(x,t)$$

for x being continuously differentiable. Using Lemma 3.4 and the fact that m is continuously differentiable the map $z \mapsto m_x(x,t)z$ is continuous and strongly monotone w.r.t. z for all $(x,t) \in U_x^0 \times U_t^0$. So $m_x(x,t)$ is non-singular and the inverse $m_x^{-1}(x,t)$ is continuous in $U_x^0 \times U_t^0$ because of Lemma 3.10.
Then (3.10) can be reformulated as

$$x' = m_x(x,t)^{-1}\left(\widetilde{f}(x,t) - m_t(x,t)\right) =: \widetilde{m}(x,t)$$

for $t \in U_t^0$ and initial value $x(t_0) = x_0 \in \mathbb{R}^{n_x}$. The function $\widetilde{m}$ is continuous as a combination of continuous functions on $U_x^0 \times U_t^0$. There is a compact set

$$Q_{\widetilde{m}} := \{(x,t) \in \mathbb{R}^{n_x} \times \mathcal{I} \mid \|x - x_0\| \leq a, \ t \in [t_0, t_0 + b]\} \subseteq U_x^0 \times U_t^0$$

for some $a, b > 0$. Hence we can apply the Peano Theorem, cf. [Zei86, Theorem 3.B]. So there exists $\tau > 0$ such that there is a solution $x_* \in C^1(\mathcal{I}_*, U_x^0)$ on the interval $\mathcal{I}_* := [t_0, t_0 + \tau] \subseteq U_t^0$. x_* solves (3.10) and with $y_*(t) := \psi(x_*(t), t) \in C(\mathcal{I}_*, \mathbb{R}^{n_y})$ we obtain a solution $(x_*, y_*) \in C^1(\mathcal{I}_*, \mathbb{R}^{n_x}) \times C(\mathcal{I}_*, \mathbb{R}^{n_y})$ of (3.9).

Uniqueness. The uniqueness proof follows the same lines as the uniqueness proof of Theorem 3.15. □

Furthermore a perturbation result can be obtained ensuring that system (3.7) has Perturbation Index 1. We concentrate on the global assumptions of Theorem 3.15.

Theorem 3.17 (Perturbation result).
Let $\mathcal{I} \subseteq \mathbb{R}$ be a compact interval. Consider the system

$$\frac{\mathrm{d}}{\mathrm{d}t}m(x,t) = f(x, y, t, \delta_x(t)) \tag{3.11a}$$

$$0 = g(x, y, t, \delta_y(t)) \tag{3.11b}$$

with functions $f \in C(\mathbb{R}^{n_x} \times \mathbb{R}^{n_y} \times \mathcal{I} \times \mathbb{R}^{n_x}, \mathbb{R}^{n_x})$, $g \in C(\mathbb{R}^{n_x} \times \mathbb{R}^{n_y} \times \mathcal{I} \times \mathbb{R}^{n_y}, \mathbb{R}^{n_y})$, $m \in C^1(\mathbb{R}^{n_x} \times \mathcal{I}, \mathbb{R}^{n_x})$ and perturbations $\delta_x \in C(\mathcal{I}, \mathbb{R}^{n_x})$ and $\delta_y \in C(\mathcal{I}, \mathbb{R}^{n_y})$. If

(i) m is strongly monotone w.r.t. x,

(ii) f is Lipschitz continuous w.r.t. x, y and δ_x,

(iii) g is Lipschitz continuous w.r.t. x, δ_y and strongly monotone w.r.t. y

then (3.11) has a unique solution $(x_*^\delta, y_*^\delta) \in C^1(\mathcal{I}, \mathbb{R}^{n_x}) \times C(\mathcal{I}, \mathbb{R}^{n_y})$ for every initial value $x_*^\delta(t_0) = x_0^\delta \in \mathbb{R}^{n_x}$. If (x_*, y_*) is the solution for $(\delta_x, \delta_y) = 0$ and $x_*(t_0) = x_0$ and if $\left\| x_0 - x_0^\delta \right\|$ is sufficiently small, then there is a constant $c > 0$ such that

$$\left\| x_* - x_*^\delta \right\|_\infty + \left\| y_* - y_*^\delta \right\|_\infty \leq c \left(\left\| x_0 - x_0^\delta \right\| + \left\| \delta_x \right\|_\infty + \left\| \delta_y \right\|_\infty \right).$$

PROOF:
Let δ_x and δ_y be given. Then the unique solvability follows directly from Theorem 3.15 because the perturbations δ_x and δ_y depend only on t. Concerning the perturbation estimate the algebraic part (3.11b) can be solved uniquely with a continuous solution function ψ with $y = \psi(x, t, \delta_y)$. This is due to Lemma 3.7. ψ is Lipschitz continuous w.r.t. x and δ_y. So with Theorem 3.15 system (3.11) is uniquely solvable for a given initial value $x_0^\delta \in \mathbb{R}^{n_x}$. The solution is denoted by (x_*^δ, y_*^δ), and for $(\delta_x, \delta_y) = 0$ and $x_*(t_0) = x_0$ we denote the solution by (x_*, y_*). We obtain for $t \in \mathcal{I}$:

$$y_*(t) = \psi(x_*(t), t, 0), \quad y_*^\delta(t) = \psi(x_*^\delta(t), t, \delta_y(t))$$

So we see that

$$\left\| y_*^\delta(t) - y_*(t) \right\| \leq c_1 \left(\left\| x_*^\delta(t) - x_*(t) \right\| + \left\| \delta_y(t) \right\| \right) \tag{3.12}$$

for a constant $c_1 > 0$. By integrating the dynamical part we see that

$$m(x_*(t), t) = m(x_0, t_0) + \int_{t_0}^t f(x_*(s), y_*(s), s, 0) \mathrm{d}s,$$

$$m(x_*^\delta(t), t) = m(x_0^\delta, t_0) + \int_{t_0}^t f(x_*^\delta(s), y_*^\delta(s), s, \delta_x(s)) \mathrm{d}s.$$

Making use of the strong monotonicity of m we observe

$$
\begin{aligned}
&\left\| x_*^\delta(t) - x_*(t) \right\| \\
\leq\ & \frac{1}{\mu} \left\| m(x_*^\delta(t), t) - m(x_*(t), t) \right\| \\
\leq\ & \frac{1}{\mu} \left\| m(x_0^\delta, t_0) - m(x_0, t_0) \right\| \\
& + \frac{1}{\mu} \int_{t_0}^t \left\| f(x_*^\delta(s), y_*^\delta(s), s, \delta_x(s)) - f(x_*(s), y_*(s), s, 0) \right\| \mathrm{d}s \\
\leq\ & \frac{1}{\mu} \left\| m(x_0^\delta, t_0) - m(x_0, t_0) \right\| \\
& + c_2 \int_{t_0}^t \left\| x_*^\delta(s) - x_*(s) \right\| + \left\| y_*^\delta(s) - y_*(s) \right\| + \left\| \delta_x(s) \right\| \mathrm{d}s \\
\overset{(3.12)}{\leq}\ & c_3 \int_{t_0}^t \left\| x_*^\delta(s) - x_*(s) \right\| \mathrm{d}s + c_4 \left(\left\| \delta_x \right\|_\infty + \left\| \delta_y \right\|_\infty + \left\| m(x_0^\delta, t_0) - m(x_0, t_0) \right\| \right)
\end{aligned}
$$

with constants $c_2, c_3, c_4 > 0$. Here we used the Lipschitz continuity of f and the fact that the perturbations δ_x and δ_y are continuous. Additionally an application of the mean value theorem gives

$$
\left\| m(x_0^\delta, t_0) - m(x_0, t_0) \right\| \leq c_5 \left\| m_x(\xi, t_0) \right\|_* \left\| x_0^\delta - x_0 \right\|
$$

for a constant $c_5 > 0$ and $\xi \in \mathbb{R}^{n_x}$. For $\left\| x_0^\delta - x_0 \right\|$ being sufficiently small $\left\| m_x(\xi, t_0) \right\|_*$ is bounded. An application of the Gronwall Lemma gives

$$
\left\| x_*^\delta(t) - x_*(t) \right\| \leq c_6 \left(\left\| \delta_x \right\|_\infty + \left\| \delta_y \right\|_\infty + \left\| x_0^\delta - x_0 \right\| \right)
$$

for a constant $c_6 > 0$. In combination with (3.12) we achieve the desired result. $\qquad\square$

Remark 3.18.
Theorem 3.15 still holds if the strong monotonicity of g w.r.t. y is replaced by the condition that there is a continuous solution function $\psi : \mathbb{R}^{n_x} \times \mathcal{I} \to \mathbb{R}^{n_y}$ which is Lipschitz continuous in x. The perturbation result stated in Theorem 3.17 can also be applied. The same is true for Theorem 3.16, if the local strong monotonicity of g is replaced by the existence of a continuous solution function in the neighborhood of the initial value (x_0, y_0, t_0) which is locally Lipschitz continuous in x. These generalizations will be important when applying the solvability results from before to the equations of the MNA in the next section.

With the decoupling presented in Lemma 3.14 the solvability and perturbation results above can be transferred to a DAE of the form (3.4). We conclude this section with two corollaries.

Corollary 3.19.
Consider a DAE of the form (3.4) on the interval $\mathcal{I} := [t_0, T]$, i.e.

$$A\frac{\mathrm{d}}{\mathrm{d}t}d(z,t) + b(z,t) = 0 \quad t \in \mathcal{I}$$

with the Assumption 3.12 and p, q *and* v, w *as in Lemma 3.13 for the matrices $d_z(z,t)$ and A^T, respectively, and initial value*

$$z_0 \in \mathcal{M}_0(t_0) = \left\{ z \in \mathbb{R}^n \mid \mathrm{q}^\top b(z, t_0) = 0 \right\}.$$

If

(i) $(x,t) \mapsto \mathrm{v}^\top A d(\mathrm{p}x, t)$ *is (locally) strongly monotone w.r.t. x (in (x_0, t_0)),*

(ii) $(x,y,t) \mapsto \mathrm{v}^\top b(\mathrm{p}x + \mathrm{q}y, t)$ *is (locally) Lipschitz continuous w.r.t. x and y (in (x_0, y_0, t_0)),*

(iii) $(x,y,t) \mapsto \mathrm{w}^\top b(\mathrm{p}x + \mathrm{q}y, t)$ *is (locally) Lipschitz continuous w.r.t. x and (locally) strongly monotone w.r.t. y (in (x_0, y_0, t_0)),*

then (3.4) has a unique continuous solution $z : \mathcal{I}_* \to \mathbb{R}^n$ with $\mathrm{p}^\top z$ being continuously differentiable on $\mathcal{I}_* = \mathcal{I}$ (on $\mathcal{I}_* = [t_0, t_0 + \tau]$ with a scalar $\tau > 0$).

PROOF:
The proof is a direct application of Theorem 3.15 (global solvability) and Theorem 3.16 (local solvability) and Lemma 3.14. □

Corollary 3.20.
Let the assumptions of Corollary 3.19 in the global sense be fulfilled. Consider additionally the perturbed system

$$A\frac{\mathrm{d}}{\mathrm{d}t}d(z,t) + b(z,t) = \delta(t), \quad t \in \mathcal{I} \tag{3.13}$$

for a perturbation $\delta \in C(\mathcal{I}, \mathbb{R}^n)$ with initial value

$$z_0^\delta \in \mathcal{M}_0^\delta(t_0) = \left\{ z \in \mathbb{R}^n \mid \mathrm{q}^\top (b(z, t_0) - \delta(t_0)) = 0 \right\}$$

and $\left\| \mathrm{pp}^\top z_0^\delta - \mathrm{pp}^\top z_0 \right\|$ being sufficiently small. Then (3.4) and (3.13) have unique solutions z_* and z_*^δ, respectively. Furthermore the following estimate holds

$$\left\| z_* - z_*^\delta \right\|_\infty \leq c \left(\left\| z_0^\delta - z_0 \right\| + \left\| \delta \right\|_\infty \right)$$

for a constant $c > 0$.

PROOF:
This is a direct consequence of Theorem 3.17 in connection with Corollary 3.19 and Lemma 3.14. □

3.3. Application in circuit simulation

In this section we describe briefly the well-known equations of the Modified Nodal Analysis (MNA) as they are the standard tool for coupled systems in circuit simulation. From an industrial point of view a suitable model for numerical simulation of electric networks has to meet two contradicting requirements. On the one hand the number of independent network variables has to be as small as possible to keep computing time sufficiently small. On the other hand the physical behavior of the electrical network should be reflected as correct as possible. A well-established modeling approach meeting these demands is based on the network's topology and finally results in the MNA equations.

In MNA the circuit's topology is modeled by a network graph consisting of nodes and branches. The physical behavior is described by the Kirchhoff circuit laws and the characteristic equations for the basic elements, namely resistors, capacitors, inductors, current and voltage sources, see [CL75, CDK87, Ria08]. We only consider elements with two contacts here, i.e. each element is represented by a branch of the network. The resulting system of equations results in a DAE, cf. [ET00, Tis96].

After having described the MNA equations we will decouple the resulting system with the decoupling presented in Lemma 3.14. Finally, we apply the solvability Theorems 3.15 and 3.16 and the perturbation Theorem 3.17. The assumptions will only be based on properties of the characteristic equations for the basic elements and on the network topology.

Basic elements and characteristics

For the basic elements of an electrical circuit, i.e. capacitors (C), resistors (R), inductors (L), current sources (I) and voltage sources (V), certain characteristic equations hold describing the physical relation between their branch currents and voltages. Formally let n_X with $X \in \{C, R, L, V, I\}$ be the number of elements of type X in the circuit. Similarly $j_X, v_X : \mathcal{I} \to \mathbb{R}^{n_X}$ are the vectors of branch currents and voltages of elements of type X. $\mathcal{I} \subseteq \mathbb{R}$ is a time interval. The constitutive relations are

$$j_C(t) = \frac{\mathrm{d}}{\mathrm{d}t} q_C(v_C(t), t), \quad j_R(t) = g_R(v_R(t), t), \quad v_L(t) = \frac{\mathrm{d}}{\mathrm{d}t} \phi_L(j_L(t), t)$$

for the charge of the capacitors, the conductance of the resistors and the flux of the inductors with functions

$$q_C : \mathbb{R}^{n_C} \times \mathcal{I} \to \mathbb{R}^{n_C}, \quad g_R : \mathbb{R}^{n_R} \times \mathcal{I} \to \mathbb{R}^{n_R} \quad \text{and} \quad \phi_L : \mathbb{R}^{n_L} \times \mathcal{I} \to \mathbb{R}^{n_L}.$$

For the sources we restrict ourselves to independent sources

$$j_I(t) = i_s(t), \quad v_V(t) = v_s(t)$$

with functions $i_s : \mathcal{I} \to \mathbb{R}^{n_I}$ and $v_s : \mathcal{I} \to \mathbb{R}^{n_V}$. For the treatment of controlled sources we refer to [ET00]. Concerning the smoothness of the functions q_C, g_R, ϕ_L, i_s and v_s we make the following assumption.

Assumption 3.21.
The element relations satisfy

$$q_C \in C^1(\mathbb{R}^{n_C} \times \mathcal{I}, \mathbb{R}^{n_C}), \quad \phi_L \in C^1(\mathbb{R}^{n_L} \times \mathcal{I}, \mathbb{R}^{n_L}), \quad g_R \in C(\mathbb{R}^{n_R} \times \mathcal{I}, \mathbb{R}^{n_R}).$$

We denote the Jacobians by

$$C(y,t) := \frac{\partial}{\partial y} q_C(y,t), \quad L(j,t) := \frac{\partial}{\partial j} \phi_L(j,t).$$

The source terms are continuous, i.e. $i_s \in C(\mathcal{I}, \mathbb{R}^{n_I})$ and $v_s \in C(\mathcal{I}, \mathbb{R}^{n_V})$.

In literature also the matrix

$$G(y,t) := \frac{\partial}{\partial y} g_R(y,t)$$

is defined. For the matrices C, L, G it is assumed that they are positive definite for all y and $t \in \mathcal{I}$. Physically this means that the elements are strictly (locally) passive, i.e. they do not emit energy, see e.g. [Tis04, Bar04]. It turns out that for applying the global solvability result from the last section we have to strengthen the passivity condition by assuming strong monotonicity. For example for the matrix $C(y,t)$ it means that

$$v_C^\top C(y,t) v_C \geq \mu_C \|v_C\|^2 \qquad \text{for all } y, v_C \in \mathbb{R}^{n_C}, \ t \in \mathcal{I}$$

due to Lemma 3.4. Although it is well-known for constant matrices that positive definiteness and strong monotonicity coincide, cf. [Saa03], the strong monotonicity condition here is important to get a uniform bound $\mu_C > 0$ from below for all $(y,t) \in \mathbb{R}^{n_C} \times \mathcal{I}$. However, the differentiability of g_R is not needed. We make the following assumption.

Assumption 3.22 (Global passivity).
The element relations q_C, ϕ_L and g_R are strongly monotone w.r.t. the first argument and g_R is Lipschitz continuous w.r.t. the first argument.

Aiming for local solvability we need the corresponding local version of Assumption 3.22.

Assumption 3.23.
The element relations q_C, ϕ_L and g_R are locally strongly monotone w.r.t. the first argument and g_R is locally Lipschitz continuous w.r.t. the first argument.

We do not require differentiability of g_R here. However, if the matrix G exists and is positive definite, then Assumption 3.23 is fulfilled. We also fix the assumption that is usually employed in the literature, cf. e.g. [ET00, Tis04, Bar04, Bod07].

Assumption 3.24 (Local passivity, [Bar04]).
The element relations q_C, ϕ_L and g_R are continuously differentiable and

$$C(y,t) := \frac{\partial}{\partial y} q_C(y,t), \quad L(j,t) := \frac{\partial}{\partial y} \phi_L(j,t), \quad G(y,t) := \frac{\partial}{\partial y} g_R(y,t).$$

are positive definite. The source terms i_s and v_s are continuous.

An example for a strongly monotone characteristic function for a diode – modeled as a nonlinear resistor – can be found in [TS02, chapter 1]. These functions are defined as a combination of exponential functions. They are in general not Lipschitz continuous on $\mathbb{R}$. Nevertheless this can be achieved by cutting off the function outside a certain area of physical interest and making a linear extension, see section 5.3 for an example.

Topology and Kirchhoff laws

An electrical network can be described topologically by an arbitrarily oriented connected graph $\mathcal{G} = (\mathcal{N}, \mathcal{B})$, where $\mathcal{N}$ is the set of nodes and $\mathcal{B}$ is the set of branches. We briefly state that a loop is a connected subgraph, where exactly two branches are incident with each node. A cutset is a subgraph of $\mathcal{G}$, such that, if removed from $\mathcal{G}$, a disconnected graph $\mathcal{G}'$ remains. But if any branch from the cutset is added to $\mathcal{G}'$ the resulting graph is connected again. We will not cover further standard graph theory here. Instead we refer to [Die05] for general graph theory and to [DK84, Tis04, Ria08, Bau12] for graph theory in the context of MNA where additional illustrating examples are given. The basic network elements – resistors, capacitors, inductors, current sources and voltage sources – are located on the branches. So it holds

$$\sum_{X \in \{C,R,L,V,I\}} n_X = |\mathcal{B}| \, .$$

The network topology is then described by the (reduced) incidence matrix

$$A \in \{-1, 0, 1\}^{(|\mathcal{N}|-1) \times |\mathcal{B}|}$$

which can be obtained by the circuit graph when eliminating an (arbitrary) reference node also called the mass node. It is given by

$$(A)_{ij} := \begin{cases} 1, & \text{if the branch } j \text{ leaves node } i, \\ -1, & \text{if the branch } j \text{ enters node } i, \\ 0, & \text{else.} \end{cases}$$

We denote by

$$j(t) = \begin{pmatrix} j_C(t) & j_R(t) & j_L(t) & j_V(t) & j_I(t) \end{pmatrix}^\top \quad \text{and}$$

$$v(t) = \begin{pmatrix} v_C(t) & v_R(t) & v_L(t) & v_V(t) & v_I(t) \end{pmatrix}^\top$$

the vectors of all branch currents and voltages for all $t \in \mathcal{I}$. The circuit modeling is based on the well-known Kirchhoff laws which hold at any time.

(i) *Kirchhoff's current law* (KCL): The algebraic sum of all branch currents having one node in common is equal to zero.

(ii) *Kirchhoff's voltage law* (KVL): The algebraic sum of voltages along any loop of the network is equal to zero.

Mathematically KCL and KVL imply

$$Aj(t) = 0 \quad \text{and} \quad A^\top v(t) = e(t)$$

where $e(t) \in \mathbb{R}^{n_e}$, $n_e := |\mathcal{N}| - 1$, $t \in \mathcal{I}$, is the vector of node potentials. The node potentials are the voltage drops at every node compared to the mass node. Furthermore the incidence matrix A can be sorted according to the branch types (w.r.t. the elements):

$$A = \begin{pmatrix} A_C & A_R & A_L & A_V & A_I \end{pmatrix}$$

The MNA equations can now be derived easily starting from the KCL by taking the following steps:

1. Apply the KCL to every node except the mass node.

2. Insert the constitutive relations for capacitors, resistors and current sources given by the functions q_C, g_R and i_s.

3. Add the constitutive relation for the inductors and insert the KVL $v_X(t) = A_X^\top e(t)$ to obtain a system formulated in the node potentials and the currents through inductors and voltage sources.

We finally arrive at the MNA equations, cf. [HRB75, CDK87]:

$$A_C \frac{\mathrm{d}}{\mathrm{d}t} q_C(A_C^\top e(t), t) + A_R g_R(A_R^\top e(t), t) + A_L j_L(t) + A_V j_V(t) + A_I i_s(t) = 0 \qquad (3.14\mathrm{a})$$

$$\frac{\mathrm{d}}{\mathrm{d}t} \phi_L(j_L(t), t) - A_L^\top e(t) = 0 \qquad (3.14\mathrm{b})$$

$$A_V^\top e(t) - v_s(t) = 0 \qquad (3.14\mathrm{c})$$

Note that the number of variables, namely $(e(t), j_L(t), j_V(t))$, has been significantly decreased compared to the starting Kirchhoff equations and element relations. Example 2.2 fits exactly into the framework of (3.14). For investigating a reasonable circuit we assume the circuit to be connected and that it is not a short circuit.

Assumption 3.25 (No short circuit, cf. [ET00]).
There are neither loops of voltage sources only nor cutsets of current sources only, i.e. the matrices

$$A_V \quad \text{and} \quad \begin{pmatrix} A_C & A_R & A_L & A_V \end{pmatrix}^\top \qquad (3.15)$$

have full column rank.

For applying the solvability results of the last section to the MNA equations we need stricter topological conditions. We make the following assumption.

Assumption 3.26.
The matrices

$$Q_C^\top A_V \quad and \quad \begin{pmatrix} A_C & A_R & A_V \end{pmatrix}^\top \tag{3.16}$$

have full column rank where Q_C is a projector with $\operatorname{im} Q_C = \ker A_C^\top$.

Topologically this means that there are neither loops of capacitors and voltage sources with at least one voltage source nor cutsets of inductors and current sources. Note that (3.16) already implies (3.15). Assumption 3.26 is also known as the topological index 1 conditions, cf. [ET00].

The MNA equations (3.14) were already classified in terms of the Tractability Index and it was shown that the index conditions can be completely given in terms of topological conditions of the circuit.

Theorem 3.27 (cf. [Tis99, ET00]).
Let Assumptions 3.24 and 3.25 be fulfilled. Then the MNA equations (3.14) represent a DAE (3.4) with a properly stated leading term. The DAE has

- *Tractability Index 0 if and only if there are no voltage sources in the circuit and the circuit has a tree containing capacitors only,*

- *Tractability Index 1 if and only if there is at least one voltage source in the circuit or there is no tree containing capacitors only and if there is neither an LI-cutset nor a CV-loop with at least one voltage sources,*

- *Tractability Index 2 otherwise.*

An LI-cutset is a cutset consisting only of inductors and current sources and a CV-loop is a loop consisting only of capacitors and voltage sources. Local solvability and perturbation results exist, cf. [Tis99, Tis04]. In the next section we investigate the solvability of the MNA equations under the Assumptions 3.21, 3.22, 3.23, 3.26. Especially a global solvability result will be obtained.

Global solvability of the MNA equations

We will now introduce the matrices for decoupling the MNA equations (3.14). Therefore let $k_{\overline{C}} := \dim(\ker A_C^\top)$ and $q_C \in \mathbb{R}^{n_e \times k_{\overline{C}}}$ whose columns form an orthonormal basis of $\ker A_C^\top$, cf. Lemma 3.13. Let then $k_C := n_e - k_{\overline{C}}$ and $p_C \in \mathbb{R}^{n_e \times k_C}$ whose columns form an orthonormal basis of $(\ker A_C^\top)^\perp$. We have $A_V^\top q_C \in \mathbb{R}^{n_V \times k_{\overline{C}}}$ and let be $q_{CV} \in \mathbb{R}^{k_{\overline{C}} \times k_{\overline{C}V}}$ whose columns form an orthonormal basis of $\ker A_V^\top q_C$, $k_{\overline{C}V} := \dim(\ker A_V^\top q_C)$. The columns of the corresponding matrix $p_{CV} \in \mathbb{R}^{k_{\overline{C}} \times k_{\overline{C}V}}$ form an orthonormal basis of $(\ker A_V^\top q_C)^\perp$ and $k_{\overline{C}V} := k_{\overline{C}} - k_{\overline{C}V}$.

Lemma 3.28.

Let Assumptions 3.26 be fulfilled. Then the matrices

$$q_C^\top A_V \quad and \quad A_R^\top q_C q_{CV} \tag{3.17}$$

have full column rank.

PROOF:

$Q_C := q_C q_C^\top$ is a projector onto $\ker A_C^\top$. With Assumption 3.26 it follows that

$$\ker\left(q_C q_C^\top A_V\right) = \{0\}.$$

Since q_C has full column rank we deduce that $\ker(q_C^\top A_V) = \{0\}$. Furthermore q_{CV} and $\begin{pmatrix} A_C & A_R & A_V \end{pmatrix}^\top$ have full column rank due to Assumption 3.26. So $q_C q_{CV}$ and $\begin{pmatrix} A_C & A_R & A_V \end{pmatrix}^\top q_C q_{CV}$ have full column rank. Since

$$\begin{pmatrix} A_C & A_R & A_V \end{pmatrix}^\top q_C q_{CV} = \begin{pmatrix} A_C^\top q_C q_{CV} \\ A_R^\top q_C q_{CV} \\ A_V^\top q_C q_{CV} \end{pmatrix} = \begin{pmatrix} 0 \\ A_R^\top q_C q_{CV} \\ 0 \end{pmatrix}$$

also $A_R^\top q_C q_{CV}$ has full column rank. □

In the following we omit the t-dependencies of the variables $e(t)$, $j_L(t)$, $j_V(t)$ to make the notation shorter. With the matrices p_C, q_C, p_{CV} and q_{CV} we can split the vector $e \in \mathbb{R}^{n_e}$. Therefore we define

$$e_C := p_C^\top e \in \mathbb{R}^{k_C}, \qquad\qquad e_{\overline{C}} := q_C^\top e \in \mathbb{R}^{k_{\overline{C}}},$$
$$e_{\overline{C}V} := p_{CV}^\top e_{\overline{C}} \in \mathbb{R}^{k_{\overline{C}V}}, \qquad\qquad e_{\overline{C}\overline{V}} := q_{CV}^\top e_{\overline{C}} \in \mathbb{R}^{k_{\overline{C}\overline{V}}},$$

compare Lemma 3.13. We can write

$$e = p_C e_C + q_C e_{\overline{C}} = p_C e_C + q_C p_{CV} e_{\overline{C}V} + q_C q_{CV} e_{\overline{C}\overline{V}}. \tag{3.18}$$

With the matrices p_C, q_C, p_{CV} and q_{CV} we can also split the MNA equations (3.14).

$$p_C^\top \left[A_C \frac{\mathrm{d}}{\mathrm{d}t} q_C(A_C^\top p_C e_C, t) + A_R g_R(A_R^\top e, t) + A_L j_L + A_V j_V + A_I i_s(t) \right] = 0 \tag{3.19a}$$

$$p_{CV}^\top q_C^\top \left[A_R g_R(A_R^\top e, t) + A_L j_L + A_V j_V + A_I i_s(t) \right] = 0 \tag{3.19b}$$

$$q_{CV}^\top q_C^\top \left[A_R g_R(A_R^\top e, t) + A_L j_L + A_I i_s(t) \right] = 0 \tag{3.19c}$$

$$\frac{\mathrm{d}}{\mathrm{d}t} \phi_L(j_L, t) - A_L^\top e = 0 \tag{3.19d}$$

$$A_V^\top e - v_s(t) = 0 \tag{3.19e}$$

Note that $A_C^\mathsf{T} q_C = 0$ and $A_V^\mathsf{T} q_C q_{CV} = 0$ due to the choice of q_C, q_{CV}. Define

$$
\begin{aligned}
A_{\overline{C}X} &:= q_C^\mathsf{T} A_X, \quad X \in \{R, L, V, I\}, \\
A_{\overline{CV}X} &:= q_{CV}^\mathsf{T} q_C^\mathsf{T} A_X, \quad X \in \{R, L, I\}.
\end{aligned}
$$

for a more compact notation. We then obtain the following lemma.

Lemma 3.29.
Let Assumptions 3.21 and 3.26 be fulfilled. The MNA equations (3.14) can be reformulated as a DAE of the form

$$
\frac{\mathrm{d}}{\mathrm{d}t} m\left(\begin{pmatrix} e_C \\ j_L \end{pmatrix}, t \right) = f\left(\begin{pmatrix} e_C \\ j_L \end{pmatrix}, e_{\overline{CV}}, t \right) \tag{3.20a}
$$

$$
0 = g\left(\begin{pmatrix} e_C \\ j_L \end{pmatrix}, e_{\overline{CV}}, t \right) \tag{3.20b}
$$

$$
\begin{pmatrix} e_{\overline{CV}} \\ j_V \end{pmatrix} = h\left(\begin{pmatrix} e_C \\ j_L \end{pmatrix}, e_{\overline{CV}}, t \right) \tag{3.20c}
$$

with

$$
\begin{aligned}
h_e(e_C, t) &:= (A_{\overline{CV}}^\mathsf{T} p_{CV})^{-1}(v_s(t) - A_V^\mathsf{T} p_C e_C) \\
H_e(e_C, t) &:= A_R^\mathsf{T} p_C e_C + A_{\overline{C}R}^\mathsf{T} p_{CV} h_e(e_C, t), \\
h_g(e_C, e_{\overline{CV}}, t) &:= g_R(A_{\overline{CV}R}^\mathsf{T} e_{\overline{CV}} + H_e(e_C, t), t) \\
h_i(e_C, j_L, e_{\overline{CV}}, t) &:= -(p_{CV}^\mathsf{T} A_{\overline{CV}})^{-1} p_{CV}^\mathsf{T} [A_{\overline{C}L} j_L + A_{\overline{C}R} h_g(e_C, e_{\overline{CV}}, t) + A_{\overline{C}I} i_s(t)]
\end{aligned}
$$

regarding the splitting (3.18) of the potentials e and

$$
\begin{aligned}
m\left(\begin{pmatrix} e_C \\ j_L \end{pmatrix}, t \right) &:= \begin{pmatrix} p_C^\mathsf{T} A_C q_C(A_C^\mathsf{T} p_C e_C, t) \\ \phi_L(j_L, t) \end{pmatrix}, \\
f\left(\begin{pmatrix} e_C \\ j_L \end{pmatrix}, e_{\overline{CV}}, t \right) &:= \begin{pmatrix} -p_C^\mathsf{T} [A_R h_g + A_L j_L + A_V h_i + A_I i_s(t)] \\ A_L^\mathsf{T} p_C e_C + A_{\overline{C}L}^\mathsf{T} p_{CV} h_e + A_{\overline{CV}L}^\mathsf{T} e_{\overline{CV}} \end{pmatrix}, \\
g\left(\begin{pmatrix} e_C \\ j_L \end{pmatrix}, e_{\overline{CV}}, t \right) &:= A_{\overline{CV}R} h_g + A_{\overline{CV}L} j_L + A_{\overline{CV}I} i_s(t), \\
h\left(\begin{pmatrix} e_C \\ j_L \end{pmatrix}, e_{\overline{CV}}, t \right) &:= \begin{pmatrix} h_e \\ h_i \end{pmatrix}.
\end{aligned}
$$

PROOF:
Using the splitting of the potentials e from (3.18) we can rewrite equations (3.19b), (3.19c) and (3.19e). First, we rewrite (3.19c) and get

$$
A_{\overline{CV}R} g_R(A_{\overline{CV}R}^\mathsf{T} e_{\overline{CV}} + A_R^\mathsf{T} p_C e_C + A_{\overline{C}R}^\mathsf{T} p_{CV} e_{\overline{CV}}) + A_{\overline{CV}L} j_L + A_{\overline{CV}I} i_s(t) = 0. \tag{3.21}
$$

Second, (3.19e) is reformulated to

$$A_V^\top \mathsf{p}_C e_C + A_{\overline{C}V}^\top \mathsf{p}_{CV} e_{\overline{C}V} - v_s(t) = 0.$$

From Lemma 3.28 we know that $A_{\overline{C}V}^\top \in \mathbb{R}^{n_V \times k_{\overline{C}}}$ has full row rank because of Assumption 3.26 and therefore is surjective. The matrix $A_{\overline{C}V}^\top \mathsf{p}_{CV} \in \mathbb{R}^{n_V \times k_{\overline{C}V}}$ is invertible. This can be seen as follows. If $w \in \ker A_{\overline{C}V}^\top \mathsf{p}_{CV}$ we have $\mathsf{p}_{CV} w = 0$ because $\mathsf{p}_{CV} w \in (\ker A_{\overline{C}V}^\top)^\perp$. The columns of p_{CV} are linearly independent and so we have that $w = 0$, showing the injectivity of $A_{\overline{C}V}^\top \mathsf{p}_{CV}$. For surjectivity let $y \in \mathbb{R}^{n_V}$. There exists $w \in \mathbb{R}^{k_{\overline{C}}}$ such that $A_{\overline{C}V}^\top w = y$ because $A_{\overline{C}V}^\top$ is surjective. Decomposing

$$w = \mathsf{p}_{CV}\mathsf{p}_{CV}^\top w + \mathsf{q}_{CV}\mathsf{q}_{CV}^\top w$$

we observe that

$$y = A_{\overline{C}V}^\top w = A_{\overline{C}V}^\top \mathsf{p}_{CV}\mathsf{p}_{CV}^\top w + A_{\overline{C}V}^\top \mathsf{q}_{CV}\mathsf{q}_{CV}^\top w = A_{\overline{C}V}^\top \mathsf{p}_{CV}\widetilde{w}$$

with $\widetilde{w} = \mathsf{p}_{CV}^\top w \in \mathbb{R}^{k_{\overline{C}V}}$ because $A_{\overline{C}V}^\top \mathsf{q}_{CV} = 0$. This also implies that $n_V = k_{\overline{C}}$ and we obtain

$$e_{\overline{C}V} = (A_{\overline{C}V}^\top \mathsf{p}_{CV})^{-1}(v_s(t) - A_V^\top \mathsf{p}_C e_C) = h_e(e_C, t). \qquad (3.22)$$

Similarly we can restate (3.19b):

$$\mathsf{p}_{CV}^\top A_{\overline{C}R} g_R(A_R^\top \mathsf{p}_C e_C + A_{\overline{C}R}^\top \mathsf{p}_{CV} e_{\overline{C}V} + A_{\overline{C}V R}^\top e_{\overline{C}V}, t)$$
$$+ \mathsf{p}_{CV}^\top A_{\overline{C}L} j_L + \mathsf{p}_{CV}^\top A_{\overline{C}V} j_V + \mathsf{p}_{CV}^\top A_{\overline{C}I} i_s(t) = 0$$

With the definition of H_e and h_g we achieve

$$\begin{aligned} j_V &= -(\mathsf{p}_{CV}^\top A_{\overline{C}V})^{-1}\mathsf{p}_{CV}^\top [A_{\overline{C}L} j_L + A_{\overline{C}R} h_g(e_C, e_{\overline{C}V}, t) + A_{\overline{C}I} i_s(t)] \\ &= h_i(e_C, j_L, e_{\overline{C}V}, t) \end{aligned}$$

because $\mathsf{p}_{CV}^\top A_{\overline{C}V}$ is invertible. So the definitions of g and h are obvious with (3.21) and (3.22). Hence equations (3.19b), (3.19c) and (3.19e) are reformulated to (3.20b) and (3.20c) with the given notation. Equations (3.19a), (3.19d) can be easily reformulated to (3.20a). $\qquad \square$

Theorem 3.30 (Global Solvability).
Consider the MNA equations (3.14) on a fixed interval $\mathcal{I} := [t_0, T]$. Let the Assumptions 3.21, 3.22 and 3.26 hold. Then the equations (3.14) have a unique solution $(e, j_L, j_V) \in C(\mathcal{I}, \mathbb{R}^{n_e + n_L + n_V})$ where $(\mathsf{p}_C^\top e, j_L) \in C^1(\mathcal{I}, \mathbb{R}^{k_C + n_L})$ for any given initial value $(\mathsf{p}_C^\top e(t_0), j_L(t_0)) = (\mathsf{p}_C^\top e_0, j_{L0})$.

PROOF:

Due to Lemma 3.29 the MNA equations (3.14) can be reduced to the form (3.20). It can be seen that a solution to (3.14) is a solution to (3.20) and vice versa regarding the splitting (3.18). For applying Theorem 3.15 to (3.20) we have to check its assumptions. Therefore we set

$$x = \begin{pmatrix} x_1 \\ x_2 \end{pmatrix} = \begin{pmatrix} e_C \\ j_L \end{pmatrix} \quad \text{and } y = e_{\overline{C}\overline{V}}.$$

(i) Obviously $m \in C^1(\mathbb{R}^{k_C+n_L} \times \mathcal{I}, \mathbb{R}^{k_C+n_L})$ because q_C, ϕ_L are continuously differentiable. So we have to check the strong monotonicity of

$$m(x,t) = \begin{pmatrix} \mathrm{p}_C^\top A_C q_C(A_C^\top \mathrm{p}_C x_1, t) \\ \phi_L(x_2, t) \end{pmatrix}$$

w.r.t. x. Let $x, \bar{x} \in \mathbb{R}^{k_C+n_L}$ then

$$
\begin{aligned}
(x - \bar{x} | m(x,t) - m(\bar{x},t)) &= (x_1 - \bar{x}_1 | \mathrm{p}_C^\top A_C q_C(A_C^\top \mathrm{p}_C x_1, t) - q_C(A_C^\top \mathrm{p}_C \bar{x}_1, t)) \\
&\quad + (x_2 - \bar{x}_2 | \phi_L(x_2, t) - \phi_L(\bar{x}_2, t)) \\
&= (A_C^\top \mathrm{p}_C x_1 - A_C^\top \mathrm{p}_C \bar{x}_1 | q_C(A_C^\top \mathrm{p}_C x_1, t) - q_C(A_C^\top \mathrm{p}_C \bar{x}_1, t)) \\
&\quad + (x_2 - \bar{x}_2 | \phi_L(x_2, t) - \phi_L(\bar{x}_2, t)) \\
&\geq \mu_C \left\| A_C^\top \mathrm{p}_C (x_1 - \bar{x}_1) \right\|^2 + \mu_L \|x_2 - \bar{x}_2\|^2 \\
&\geq \frac{\mu_C}{\left\| (A_C^\top \mathrm{p}_C)^+ \right\|_*^2} \|x_1 - \bar{x}_1\|^2 + \mu_L \|x_2 - \bar{x}_2\|^2 \\
&\geq \mu \|x - \bar{x}\|^2
\end{aligned}
$$

for constants $\mu_C, \mu_L, \mu > 0$. Here we used the strong monotonicity of q_C, ϕ_L and the fact that $A_C^\top \mathrm{p}_C$ has full column rank because of the choice of p_C. We also used

$$I_{n_C} = (A_C^\top \mathrm{p}_C)^+ (A_C^\top \mathrm{p}_C)$$

where $(A_C^\top \mathrm{p}_C)^+$ is the Moore-Penrose inverse of $A_C^\top \mathrm{p}_C$, cf. [BG03].

(ii) For showing that

$$f(x,y,t) := \begin{pmatrix} -\mathrm{p}_C^\top (A_R h_g(x_1, y, t) + A_L x_2 + A_V h_i(x_1, x_2, y, t) + A_I i_s(t)) \\ A_L^\top \mathrm{p}_C x_1 + A_{\overline{C}L}^\top \mathrm{p}_{CV} h_e(x_1, t) + A_{\overline{C}\overline{V}L}^\top y \end{pmatrix}$$

is continuous and Lipschitz continuous w.r.t. x and y, we have to recall the definition of the auxiliary functions h_e, H_e, h_g and h_i of Lemma 3.29. They are continuous and they are Lipschitz continuous w.r.t. e_C, j_L and $e_{\overline{C}\overline{V}}$ as a combination of Lipschitz continuous functions. Consequently f is continuous on $\mathbb{R}^{k_C+n_L} \times \mathbb{R}^{k_{\overline{C}\overline{V}}} \times \mathcal{I}$ and also Lipschitz continuous w.r.t. x and y. Note that we have assumed v_s and i_s to be continuous (Assumption

3.21) and g_R to be Lipschitz continuous w.r.t. the first argument (Assumption 3.22).
(iii) The function

$$g(x, y, t) = \left(A_{\overline{CV}R} h_g(x_1, y, t) + A_{\overline{CV}L} x_2 + A_{\overline{CV}I} i_s(t) \right)$$

is continuous on $\mathbb{R}^{k_C + n_L} \times \mathbb{R}^{k_{\overline{CV}}} \times \mathcal{I}$ and Lipschitz continuous w.r.t. x with the same arguments as in (ii). The strong monotonicity w.r.t. y has to be checked. Let $y, \bar{y} \in \mathbb{R}^{k_{\overline{CV}}}$ then we observe with the strong monotonicity of g_R:

$$
\begin{aligned}
&(y - \bar{y} \mid g(x, y, t) - g(x, \bar{y}, t)) \\
=\ &(y - \bar{y} \mid A_{\overline{CV}R}(g_R(A_{\overline{CV}R}^{\mathsf{T}} y + H_e(e_C, t), t) - g_R(A_{\overline{CV}R}^{\mathsf{T}} \bar{y} + H_e(e_C, t), t))) \\
=\ &(A_{\overline{CV}R}^{\mathsf{T}}(y - \bar{y}) \mid g_R(A_{\overline{CV}R}^{\mathsf{T}} y + H_e(e_C, t), t) - g_R(A_{\overline{CV}R}^{\mathsf{T}} \bar{y} + H_e(e_C, t), t)) \\
\geq\ &\mu_R \left\| A_{\overline{CV}R}^{\mathsf{T}}(y - \bar{y}) \right\|^2 \\
\geq\ &\frac{\mu_R}{\left\| (A_{\overline{CV}R}^{\mathsf{T}})^+ \right\|_*^2} \left\| y - \bar{y} \right\|^2 \\
=\ &\tilde{\mu} \left\| y - \bar{y} \right\|^2
\end{aligned}
$$

for constants $\mu, \tilde{\mu} > 0$. Here $A_{\overline{CV}R}^{\mathsf{T}}$ has full column rank due to Lemma 3.28 and therefore we used the Moore-Penrose inverse $(A_{\overline{CV}R}^{\mathsf{T}})^+$ as before. We conclude that the assumptions of Theorem 3.15 are fulfilled and there exists a unique solution for the reduced decoupled MNA equations (3.20). With $e_{\overline{CV}}$ and j_V given by (3.20c) and the splitting (3.18) we find the unique solution of the MNA equations (3.14) for the given initial value. □

Theorem 3.31 (Local Solvability).
Consider the MNA equations (3.14) on a fixed interval $\mathcal{I} := [t_0, T]$. Let the Assumptions 3.21, 3.23 and 3.26 hold. Furthermore let an inital value $z_0 := (e_0, j_{L0}, j_{V0})$ be given which fulfills equations (3.19b), (3.19c) and (3.19e). Then there exists a scalar $\tau > 0$ such that the MNA equations (3.14) have a unique solution $(e, j_L, j_V) \in C(\mathcal{I}_*, \mathbb{R}^{n_e + n_L + n_V})$ on $\mathcal{I}_* := [t_0, t_0 + \tau] \subseteq \mathcal{I}$ where $(p_C^{\mathsf{T}} e, j_L) \in C^1(\mathcal{I}_*, \mathbb{R}^{k_C + n_L})$.

PROOF:
We use the same arguments as in the Proof of Theorem 3.30 and apply Theorem 3.16 instead of Theorem 3.15 using the decoupling of the MNA from Lemma 3.29. The initial value z_0 is chosen such that

$$(x_0, y_0) := ((p_C^{\mathsf{T}} e_0, j_{L0})^{\mathsf{T}}, q_{CV}^{\mathsf{T}} q_C^{\mathsf{T}} e_0) \in \mathcal{M}_0(t_0)$$

with $\mathcal{M}_0$ being defined as in Theorem 3.16. So after applying Theorem 3.16 we again apply the same arguments as in the proof of Theorem 3.30. So there exists $\tau > 0$ and $\mathcal{I}_* := [t_0, t_0 + \tau] \subseteq \mathcal{I}$ with a unique (local) solution for the MNA equations (3.14). □

We conclude this chapter by showing a perturbation result for the MNA equations (3.14) considering the case of global assumptions.

Theorem 3.32 (Perturbation result).
Let Assumptions 3.21, 3.22 and 3.26 hold. Consider the perturbed MNA equations

$$A_C \frac{\mathrm{d}}{\mathrm{d}t} q_C(A_C^\top e(t), t) + A_R g_R(A_R^\top e(t), t) + A_L j_L(t) + A_V j_V(t) + A_I i_s(t) = \delta_e(t) \quad (3.23\text{a})$$

$$\frac{\mathrm{d}}{\mathrm{d}t} \phi_L(j_L(t), t) - A_L^\top e(t) = \delta_L(t) \quad (3.23\text{b})$$

$$A_V^\top e(t) - v_s(t) = \delta_V(t) \quad (3.23\text{c})$$

with perturbations $\delta_e \in C(\mathcal{I}, \mathbb{R}^{n_e})$, $\delta_L \in C(\mathcal{I}, \mathbb{R}^{n_L})$ and $\delta_V \in C(\mathcal{I}, \mathbb{R}^{n_V})$. Then the equations (3.23) have a unique solution

$$(e^\delta, j_L^\delta, j_V^\delta) \in C(\mathcal{I}, \mathbb{R}^{n_e + n_L + n_V}) \quad \text{where} \quad (\mathrm{p}_C^\top e^\delta, j_L^\delta) \in C^1(\mathcal{I}, \mathbb{R}^{k_C + n_L})$$

for any given initial value $(\mathrm{p}_C^\top e^\delta(t_0), j_L^\delta(t_0))^\top = (\mathrm{p}_C^\top e_0^\delta, j_{L0}^\delta)^\top =: x_0^\delta$. Let (e, j_L, j_V) be the solution for $(\delta_e, \delta_L, \delta_V) = 0$ with initial value $(\mathrm{p}_C^\top e(t_0), j_L(t_0))^\top = (\mathrm{p}_C^\top e_0, j_{L0})^\top =: x_0$. If $\|x_0 - x_0^\delta\|$ is sufficiently small then there is a $c > 0$ such that

$$\left\| e - e^\delta \right\|_\infty + \left\| j_L - j_L^\delta \right\|_\infty + \left\| j_V - j_V^\delta \right\|_\infty \le c \left(\left\| x_0 - x_0^\delta \right\| + \left\| \delta_e \right\|_\infty + \left\| \delta_L \right\|_\infty + \left\| \delta_V \right\|_\infty \right).$$

PROOF:
We set $\delta = (\delta_e, \delta_L, \delta_V)^\top$ and apply the same decomposition as in Lemma 3.29 and obtain perturbed functions

$$
\begin{aligned}
h_e^\delta(e_C, t, \delta(t)) &:= (A_{\overline{C}V}^\top \mathrm{p}_{CV})^{-1}(v_s(t) - A_V^\top \mathrm{p}_C e_C + \delta_V(t)), \\
H_e^\delta(e_C, t, \delta(t)) &:= A_R^\top \mathrm{p}_C e_C + A_{\overline{C}R}^\top \mathrm{p}_{CV} h_e^\delta(e_C, t, \delta(t)), \\
h_g^\delta(e_C, e_{\overline{CV}}, t, \delta(t)) &:= g_R(A_{\overline{CV}R}^\top e_{\overline{CV}} + H_e^\delta(e_C, t, \delta(t)), t)
\end{aligned}
$$

and

$$
\begin{aligned}
& h_i^\delta(e_C, j_L, e_{\overline{CV}}, t, \delta(t)) \\
& := -(\mathrm{p}_{CV}^\top A_{\overline{C}V})^{-1} \mathrm{p}_{CV}^\top \left[A_{\overline{C}L} j_L + A_{\overline{C}R} h_g^\delta(e_C, e_{\overline{CV}}, t, \delta(t)) + A_{\overline{C}I} i_s(t) - q_C^\top \delta_e(t) \right].
\end{aligned}
$$

All functions are Lipschitz continuous in δ. Furthermore we have

$$
f^\delta \left(\begin{pmatrix} e_C \\ j_L \end{pmatrix}, e_{\overline{CV}}, t, \delta(t) \right) := \begin{pmatrix} -\mathrm{p}_C^\top \left[A_R h_g^\delta + A_L j_L + A_V h_i^\delta + A_I i_s(t) - \delta_e(t) \right] \\ A_L^\top \mathrm{p}_C e_C + A_{\overline{C}L}^\top \mathrm{p}_{CV} h_e + A_{\overline{CV}L}^\top e_{\overline{CV}} + \delta_L(t) \end{pmatrix}
$$

$$
g^\delta \left(\begin{pmatrix} e_C \\ j_L \end{pmatrix}, e_{\overline{CV}}, t, \delta(t) \right) := A_{\overline{CV}R} h_g^\delta + A_{\overline{CV}L} j_L + A_{\overline{CV}I} i_s(t) - q_{CV}^\top q_C^\top \delta_e(t)
$$

$$
h^\delta \left(\begin{pmatrix} e_C \\ j_L \end{pmatrix}, e_{\overline{CV}}, t, \delta(t) \right) := \begin{pmatrix} h_e^\delta \\ h_i^\delta \end{pmatrix}
$$

and these are Lipschitz continuous w.r.t. δ as well. Therefore Theorem 3.17 with Remark 3.18 can be applied and gives the desired result. $\qquad \square$

3.4. Conclusion

In this chapter we have derived a global and a local existence result for nonlinear differential-algebraic equations of the form (3.4) having Perturbation Index 1, cf. Corollaries 3.19 and 3.20. The decoupling into algebraic and dynamical parts is based on an approach using orthonormal bases of certain subspaces which was developed by Jansen, cf. [Jan13]. The resulting algebraic part is assumed to be given by a (locally) strongly monotone function and is solved by Lemma 3.7 (Lemma 3.10), where we investigated the solvability of such equations with regard to parameters. Furthermore a strongly monotone derivative part and further assumptions such as Lipschitz continuity ensure a priori estimates and finally the existence of a global (local) unique solution and perturbation estimate, cf. Theorems 3.15, 3.16 and 3.17. The solvability results are applicable to the equations of the Modified Nodal Analysis (MNA) under the topological index 1 conditions (Assumption 3.26), cf. Theorems 3.30 and 3.31. The usual passivity assumptions for the element functions had to be slightly extended to fit into the concept of strong monotonicity. However, differentiability of the conductivity function g_R is not necessary anymore. Finally, we proved that the MNA equations with the topological index 1 condition still have Perturbation Index 1, cf. Theorem 3.32.

The foundations and ideas collected in this chapter and the concept of strong monotonicity will be important in chapters 4 and 5. Especially the decoupling of the MNA equations (Lemma 3.29) and the global solvability of the algebraic part (Theorem 3.30) will be applied in chapter 5 when investigating prototype coupled systems in circuit simulation.

It should be possible to extend the results of this chapter in various directions. For example the smoothness assumption concerning the MNA equations (Assumption 3.21) could be generalized. For example the derivative of the charge and flux functions q_C and ϕ_L could be understood in the weak sense, i.e. it exists almost everywhere. Instead of solving the inherent ODE with the Peano Theorem, an application of the more general Theorem of Carathéodory would be necessary. Thus also the assumption that the source functions i_s and v_s are continuous could be dropped, requiring the functions i_s and v_s only to be integrable. This would result in a solution where the derivatives of the dynamical components exist almost everywhere and the algebraic solution components are integrable. A further extension could be to investigate the Tractability Index 2 case, cf. Theorem 3.27. Therefore the idea of the decoupling needs to be carried further, cf. [Jan13] for details of the decoupling in the higher index setting.

4. Nonlinear ADAEs with monotone operators

In this chapter an operator approach is investigated for solving nonlinear abstract differential-algebraic equations (ADAEs) with the Galerkin approach. We study a system of the form

$$\mathcal{A}^* \left[\mathcal{D}u(t) \right]' + \mathcal{B}(t)(u(t)) = r(t) \quad \text{for almost all } t \in [t_0, T] \tag{4.1}$$

with operators $\mathcal{A}, \mathcal{D} : V \to Z$, $\mathcal{B}(t) : V \to V^*$, $t \in \mathcal{I}$, and V, Z being real Banach spaces and V^* being the dual space of V. The operators $\mathcal{A}$ and $\mathcal{D}$ are linear whereas the operators $\mathcal{B}(t)$ are nonlinear. $\mathcal{A}^*$ is the dual operator of $\mathcal{A}$ and the derivative $[\mathcal{D}u]'$ is understood in the sense of generalized derivatives in the setting of evolution triples, cf. Appendix A.3.

If $\mathcal{A}$ and $\mathcal{B}$ are identity operators (and $V = Z$) equation (4.1) is reduced to a standard evolution equation with well-known solvability results, see e.g. [Zei90b, Rou05, Eva08]. Standard approaches for treating evolution equations are the Galerkin and the semigroup approach. For the semigroup approach we refer to [FY99, FR99, RK04] and the comments in chapter 2 of this thesis.

From a numerical point of view the Galerkin approach is preferable because it automatically provides a numerical solution procedure. For ADAEs of the form (4.1) the Galerkin approach was investigated by Tischendorf in [Tis04] for linear strongly monotone operators $\mathcal{B}(t)$. In this chapter we will extend this approach to the nonlinear case in this chapter. In the Galerkin approach system (4.1) is approximated by a system on a finite dimensional subspace V_n (dim $V_n = n$) of V. This system is given by the Galerkin equations and yields a differential-algebraic equation (DAE). Having solved these, the Galerkin solutions converge to the original solution to (4.1) as $n \to \infty$ by means of a priori estimates and monotonicity arguments.

The unique solvability of the nonlinear Galerkin equations poses an interesting subproblem here. Although the resulting DAE has Index 1 character, well-known solvability results from standard DAE theory, see [GM86, KM06, LMT13] for example, cannot be applied because the required smoothness conditions are not fulfilled. Furthermore solvability needs to be ensured on a fixed time interval as otherwise it may depend on the

Galerkin step n and it is not clear what happens to the existence interval if n tends to infinity. We aim for global solvability on the fixed interval $[t_0, T]$ here.

In this chapter we extend the results obtained for the linear case of (4.1) in [Tis04, chapter 4] to the nonlinear case. So instead of a linear operator family $\mathcal{B}(t)$ we allow the operator family to be nonlinear under comparable conditions concerning the monotonicity and boundedness of $\mathcal{B}(t)$. Furthermore the structural condition $\mathcal{A} = \mathcal{D}$ from [Tis04] for the dynamical part is weakened to a condition of the form $\mathcal{A} = \mathcal{T}\mathcal{D}$ with a certain operator $\mathcal{T}$. Together with other structural conditions this is discussed in the following section. Next we will prove that the Galerkin equations corresponding to (4.1) have a unique solution. Then the main unique solvability result is presented and strong convergence of the sequence of Galerkin solutions is shown. Finally we investigate the behavior of the solution with regard to perturbations on the right hand side and the initial value and classify (4.1) in terms of the ADAE Index.

4.1. Structural assumptions

In this section we discuss the structural assumptions for solving the following ADAE:

$$\mathcal{A}^* \left[\mathcal{D}u(t)\right]' + \mathcal{B}(t)(u(t)) = r(t) \qquad \text{for almost all } t \in [t_0, T] \tag{4.2a}$$
$$\mathcal{D}u(t_0) = z_0 \tag{4.2b}$$

We set $\mathcal{I} := [t_0, T]$ with $t_0 \le T < \infty$ and let V, Z be real Banach spaces. The operators $\mathcal{D}$, $\mathcal{A} : V \to Z$ are linear, bounded and independent of time whereas the operator $\mathcal{B}$ is time dependent and depends nonlinearly on u. The derivative $[\mathcal{D}u(t)]'$ is to be understood in the generalized sense in the setting of evolution triples. With $\mathcal{A}^* : Z^* \to V^*$ we denote the dual operator of $\mathcal{A} \in L(V, Z)$, i.e. the unique operator $\mathcal{A}^* : Z^* \to V^*$ satisfying

$$\langle \mathcal{A}^* z^*, v \rangle_V = \langle z^*, \mathcal{A}v \rangle_Z \quad \forall z^* \in Z^*, \ v \in V,$$

cf. [Wer05, chapters III.4 and V.5]. Equation (4.2a) is assumed to hold for almost all (f.a.a.) $t \in \mathcal{I}$, i.e. for all $t \in \mathcal{I} \setminus U$ where U is a set of measure zero. System (4.2) fits into the framework (2.2) presented in chapter 2. More precisely we assume the following basic properties.

Assumption 4.1 (Basic properties).
Let $\mathcal{I} := [t_0, T] \subseteq \mathbb{R}$ be a fixed time interval. Let V be a real reflexive and separable Banach space and let $Z \subseteq H \subseteq Z^*$ be an evolution triple.
The mappings $\mathcal{D} : V \to Z$ and $\mathcal{A} : V \to Z$ are linear and continuous. Furthermore $\mathcal{D}$ is surjective and $r \in L_2(\mathcal{I}, V^*)$. We have $\mathcal{B}(t) : V \to V^*$ for all $t \in \mathcal{I}$.

For basics on evolution triples we refer to Appendix A.3. Equation (4.2a) is an equation in V^* and thus can be written equivalently as

$$\langle [\mathcal{D}u(t)]', \mathcal{A}v \rangle_Z + \langle \mathcal{B}(t)(u(t)), v \rangle_V = \langle r(t), v \rangle_V \quad \forall v \in V, \text{ f.a.a. } t \in \mathcal{I}.$$

The solution space $W^1_{2,\mathcal{D}}(\mathcal{I}; V, Z, H)$

Let Assumption 4.1 be fulfilled. We define the space

$$W^1_{2,\mathcal{D}}(\mathcal{I}; V, Z, H) := \left\{ u \in L_2(\mathcal{I}, V) \mid [\mathcal{D}u]' \in L_2(\mathcal{I}, Z^*) \right\}$$

which was first introduced in [Tis04] with $[\mathcal{D}u]'$ denoting the unique generalized derivative of $\mathcal{D}u$. It is given by the relation

$$\int_{\mathcal{I}} \varphi'(t)\mathcal{D}u(t)\mathrm{d}t = -\int_{\mathcal{I}} \varphi(t)\,[\mathcal{D}u]'(t)\mathrm{d}t \quad \forall \varphi \in C_0^\infty(\mathcal{I}). \tag{4.3}$$

Here $\mathcal{D}u : \mathcal{I} \to Z$ is defined by

$$(\mathcal{D}u)(t) := \mathcal{D}u(t), \quad t \in \mathcal{I}$$

and equation (4.3) is to be understood in Z^*. So $\mathcal{D}u(t) \in Z \subseteq H$ is identified with the unique representative in Z^*, cf. Proposition A.6. This means we can write equivalently

$$\int_{\mathcal{I}} \varphi'(t)(\mathcal{D}u(t) \mid z)_H \mathrm{d}t = -\int_{\mathcal{I}} \varphi(t)\langle [\mathcal{D}u]'(t), z \rangle_Z \mathrm{d}t \quad \forall \varphi \in C_0^\infty(\mathcal{I}), \ z \in Z,$$

cf. Appendix A.3. We also write $[\mathcal{D}u]'(t) = [\mathcal{D}u(t)]'$. The space $W^1_{2,\mathcal{D}}(\mathcal{I}; V, Z, H)$ is an extension of the usual solution space

$$W^1_2(\mathcal{I}; Z, H) := \left\{ u \in L_2(\mathcal{I}, Z) \mid u' \in L_2(\mathcal{I}, Z^*) \right\}$$

for evolution equations formulated on Banach spaces of the form

$$u'(t) + \mathcal{B}(t)u(t) = r(t) \quad \text{f.a.a. } t \in \mathcal{I},$$

cf. e.g. [Zei90a, chapter 23] or [Emm04, chapter 8]. We will use the shortened notation

$$\begin{aligned} W^1_2 &:= W^1_2(\mathcal{I}; Z, H), \\ W^1_{2,\mathcal{D}} &:= W^1_{2,\mathcal{D}}(\mathcal{I}; V, Z, H) \end{aligned}$$

and collect some properties of $W^1_{2,\mathcal{D}}$.

Proposition 4.2.
Let Assumption 4.1 be fulfilled. Then the following holds:

(i) The space $W_{2,\mathcal{D}}^1$ is a real Banach space with the norm

$$\|u\|_{W_{2,\mathcal{D}}^1} := \|u\|_{L_2(\mathcal{I},V)} + \left\|[\mathcal{D}u]'\right\|_{L_2(\mathcal{I},Z^*)}.$$

(ii) If $u \in W_{2,\mathcal{D}}^1$ then $\mathcal{D}u \in W_2^1$.

(iii) For every $u \in W_{2,\mathcal{D}}^1$ there exists a unique continuous function $z_{\mathcal{D}u} : \mathcal{I} \to H$ which coincides with $\mathcal{D}u$ f.a.a. $t \in \mathcal{I}$. Furthermore there is a constant $c > 0$ such that

$$\max_{t \in \mathcal{I}} \|z_{\mathcal{D}u}(t)\|_H \leq c \|\mathcal{D}u\|_{W_2^1}.$$

(iv) Let be $\mathcal{D}u, \mathcal{A}v \in W_2^1$ and s, t with $t_0 \leq s \leq t \leq T$. Then the integration by parts formula

$$(\mathcal{D}u(t)|\mathcal{A}v(t))_H - (\mathcal{D}u(s)|\mathcal{A}v(s))_H = \int_s^t \langle [\mathcal{D}u(\tau)]', \mathcal{A}v(\tau)\rangle_Z + \langle [\mathcal{A}v(\tau)]', \mathcal{D}u(\tau)\rangle_Z \mathrm{d}\tau$$

holds. The values $\mathcal{D}u(t), \mathcal{A}v(t), \mathcal{D}u(s), \mathcal{A}v(s)$ are the values of the corresponding continuous functions $z_{\mathcal{D}u}, z_{\mathcal{A}v} : \mathcal{I} \to H$ in the sense of (iii).

PROOF:
(i), (ii) This was done in [Tis04, Propositions 4.14 and 4.15].
(iii) This is a direct consequence of (ii) and the fact that the embedding $W_2^1 \subseteq C(\mathcal{I}, H)$ is continuous, cf. Proposition A.10.
(iv) An application of the integration by parts formula in Proposition A.10 reveals the desired formula. □

Proposition 4.2 (iii) justifies the following assumption concerning the initial condition of equation (4.2a).

Assumption 4.3. *(Initial condition)*
The initial value z_0 is in H. This means that we require formally

$$z_{\mathcal{D}u}(t_0) = z_0 \in H$$

in (4.2b) with $z_{\mathcal{D}u} \in C(\mathcal{I}, H)$ in the sense of Proposition 4.2 (iii).

Assumption 4.4. *(Projectors)*
The null space of $\mathcal{D}$ splits the space V, i.e. there exists a projection operator $\mathcal{Q} \in L(V)$ with

$$\mathrm{im}\, \mathcal{Q} = \ker \mathcal{D}.$$

By $\mathcal{P}$ we denote the complementary projector of $\mathcal{Q}$.

For more details on projection operators we refer to Appendix A.3.

Lemma 4.5.
Let Assumptions 4.1 and 4.4 be fulfilled. Then there exists a constant $c > 0$ such that

$$\|v\|_V \leq c \|\mathcal{D}v\|_Z \quad \forall v \in \ker \mathcal{Q}.$$

PROOF:
We remark that $\ker \mathcal{Q} = \operatorname{im} \mathcal{P}$. We show that the operator

$$\mathcal{D}|_{\ker \mathcal{Q}} : \ker \mathcal{Q} \to Z, \qquad \mathcal{D}|_{\ker \mathcal{Q}} v := \mathcal{D}v \quad \forall v \in \ker \mathcal{Q}$$

is linear, continuous and bijective. $\mathcal{D}|_{\ker \mathcal{Q}}$ is linear and continuous because $\mathcal{D}$ is. For the
surjectivity let $z \in Z$. Then there is a $v \in V$ s.t. $\mathcal{D}v = z$ because $\mathcal{D}$ is surjective. Thus
v can be written uniquely as $v = v_1 + v_2 \in \ker \mathcal{Q} \oplus \operatorname{im} \mathcal{Q}$ where $\operatorname{im} \mathcal{Q} = \ker \mathcal{D}$ because $\mathcal{Q}$
is a projection operator. Then $\mathcal{D}v_1 = \mathcal{D}v_1 + \mathcal{D}v_2 = \mathcal{D}v = z$ and so $\mathcal{D}|_{\ker \mathcal{Q}}$ is surjective.
For injectivity let $v_1, v_2 \in \ker \mathcal{Q}$ and $\mathcal{D}v_1 = \mathcal{D}v_2$. Then $(v_1 - v_2) \in \ker \mathcal{D} = \operatorname{im} \mathcal{Q}$. So
$v_1 = v_2$. Hence the inverse operator $\mathcal{D}|_{\ker \mathcal{Q}}^{-1}$ is continuous (see e.g. [Wer05], Kor. IV 3.4)
and we have

$$\|v\|_V = \left\| \mathcal{D}|_{\ker \mathcal{Q}}^{-1} \mathcal{D}|_{\ker \mathcal{Q}} v \right\|_V \leq c \|\mathcal{D}v\|_Z \quad \forall v \in \ker \mathcal{Q}$$

for a constant $c > 0$. $\qquad \square$

We define the space

$$C_{\mathcal{D}}^1(\mathcal{I}; V, Z) := \left\{ u \in C(\mathcal{I}, V) \mid \mathcal{D}u \in C^1(\mathcal{I}, Z) \right\}.$$

and introduce the norm

$$\|u\|_{C_{\mathcal{D}}^1} := \max_{t \in \mathcal{I}} \|u(t)\|_V + \max_{t \in \mathcal{I}} \left\| [\mathcal{D}u(t)]' \right\|_Z.$$

We write shortly $C_{\mathcal{D}}^1$ instead of $C_{\mathcal{D}}^1(\mathcal{I}; V, Z)$. The next lemma shows that the space $W_{2,\mathcal{D}}^1$
is an extension of the space $C_{\mathcal{D}}^1$ which we already encountered in chapter 2.

Lemma 4.6.
Let Assumptions 4.1, 4.4 be fulfilled. Then the following holds.

(i) The set of polynomials $w : \mathcal{I} \to V$, i.e.

$$w(t) = b_0 + b_1 t + b_2 t^2 + \cdots + b_n t^n$$

with $b_i \in V$ for all i and $n = 0, 1, \ldots$ is dense in $W_{2,\mathcal{D}}^1$.

(ii) $C_{\mathcal{D}}^1 \subseteq W_{2,\mathcal{D}}^1$ dense.

PROOF:
(i) Let $u \in W_{2,\mathcal{D}}^1$, then $\mathcal{D}u \in W_2^1$ with Proposition 4.2 (ii). Since the set of polynomials $z : \mathcal{I} \to Z$ with coefficients in Z is dense in $W_2^1(\mathcal{I}; Z, H)$, cf. [Zei90a, Proposition 23.23], there is a sequence (z_n) of such polynomials with

$$z_n \to \mathcal{D}u \quad \text{in } W_2^1 \quad \text{as } n \to \infty.$$

Then there is a unique $v_n(t) \in \text{im } \mathcal{P} = \ker \mathcal{Q}$ with $\mathcal{D}v_n(t) = z_n(t)$ and v_n is a polynomial with coefficients in V because $\mathcal{D}|_{\ker \mathcal{Q}}$ is bijective. Furthermore the set of polynomials with coefficients in V is dense in $L_2(\mathcal{I}, V)$ and hence there is a sequence (u_n) of such polynomials with

$$u_n \to u \quad \text{in } L_2(\mathcal{I}, V) \quad \text{as } n \to \infty.$$

We define

$$w_n := v_n + \mathcal{Q}u_n \in W_{2,\mathcal{D}}^1$$

which is a polynomial with coefficients in V. With Lemma 4.5 (continuity of $\mathcal{D}|_{\ker \mathcal{Q}}^{-1}$) we get

$$
\begin{aligned}
\|w_n - u\|_{W_{2,\mathcal{D}}^1} &= \|(\mathcal{P} + \mathcal{Q})(w_n - u)\|_{L_2(\mathcal{I},V)} + \left\| z_n' - [\mathcal{D}u]' \right\|_{L_2(\mathcal{I},Z^*)} \\
&\leq \|v_n - \mathcal{P}u)\|_{L_2(\mathcal{I},V)} + \|\mathcal{Q}u_n - \mathcal{Q}u\|_{L_2(\mathcal{I},V)} + \left\| z_n' - [\mathcal{D}u]' \right\|_{L_2(\mathcal{I},Z^*)} \\
&\leq c \left(\|z_n - \mathcal{D}u)\|_{L_2(\mathcal{I},Z)} + \|u_n - u\|_{L_2(\mathcal{I},V)} \right) + \|z_n - \mathcal{D}u\|_{W_2^1}
\end{aligned}
$$

with a $c > 0$. The right hand side tends to 0 as $n \to \infty$ and this proves (i).
(ii) Since the set of polynomials with coefficients in V is a subset of $C_\mathcal{D}^1$ we clearly have that $C_\mathcal{D}^1$ is dense in $W_{2,\mathcal{D}}^1$ with (i). $\qquad\square$

Having introduced the proper solution space we can formulate the problem to be investigated as follows:

Find a solution $u \in W_{2,\mathcal{D}}^1(\mathcal{I}; V, Z, H)$ such that for all $v \in V$ and f.a.a. $t \in \mathcal{I}$:

$$
\begin{aligned}
\langle [\mathcal{D}u(t)]', \mathcal{A}v \rangle_Z + \langle \mathcal{B}(t)(u(t)), v \rangle_V &= \langle r(t), v \rangle_V, \qquad &(4.4\text{a}) \\
\mathcal{D}u(t_0) &= z_0. \qquad &(4.4\text{b})
\end{aligned}
$$

In the next two sections we discuss a structural condition $\mathcal{A} = \mathcal{T}\mathcal{D}$ for some operator $\mathcal{T}$ and the properties of the operators $\mathcal{B}(t) : V \to V^*$.

The operator $\mathcal{T}$

The integration by parts formula, cf. Proposition 4.2 (iv), is a major key for proving unique solvability for evolution equations, see [Zei90a], [Zei90b] or [Rou05]. In order to have it applicable to (4.4) some structural conditions on the operators $\mathcal{A}$ and $\mathcal{D}$ have to be imposed.

Assumption 4.7 (Operator $\mathcal{T}$).
Let Assumption 4.1 be fulfilled and we assume that

$$\mathcal{A} = \mathcal{T}\mathcal{D}$$

where $\mathcal{T} \in L(Z)$ is bijective and the operator $\mathcal{T}_H \in L(H)$ with $\mathcal{T}_H z = \mathcal{T}z$ for all $z \in Z$ has the following properties:

(i) $\mathcal{T}_H$ is self-adjoint, i.e. $(x|\mathcal{T}_H y)_H = (\mathcal{T}_H x|y)_H \ \forall x, y \in H$.

(ii) There exists $\mathcal{S} \in L(H)$ such that $\mathcal{T}_H = \mathcal{S}^{*H}\mathcal{S}$. By $\mathcal{S}^{*H}$ we denoted the adjoint operator of $\mathcal{S}$ in the Hilbert space setting, i.e. the unique operator satisfying

$$(\mathcal{S}x|y)_H = (x|\mathcal{S}^{*H}y)_H \ \forall x, y \in H.$$

In [Tis04] the requirement $\mathcal{A} = \mathcal{D}$ is made instead of imposing Assumption 4.7 which is more general. We state some properties of the operator $\mathcal{T}$ which will be used in the later analysis.

Lemma 4.8 (Properties of $\mathcal{T}$).
Let Assumption 4.1 be fulfilled and $\mathcal{T}$ be defined as in Assumption 4.7. Then the following holds.

(i) There is a unique $\mathcal{T}_H \in L(H)$ with $\mathcal{T}_H|_Z = \mathcal{T}$.

(ii) $\mathcal{T}_H$ is bijective and $\mathcal{T}_H^{-1} = (\mathcal{T}^{-1})_H$ where $\mathcal{T}_H^{-1}$ is the unique extension of $\mathcal{T}^{-1}$ to $L(H)$.

(iii) There is a constant $c > 0$ such that $\|v\|_H \leq c \|\mathcal{S}v\|_H$ for all $v \in H$.

(iv) $\mathcal{T}_H$ is strongly monotone, i.e. there is $\mu > 0$ such that $(v|\mathcal{T}_H v)_H \geq \mu \|v\|_H^2$ for all $v \in H$.

PROOF:
(i) Since $Z \subseteq H \subseteq Z^*$ is an evolution triple we have $\|z\|_H \leq c \|z\|_Z$ for all $z \in Z$ with a fixed constant $c > 0$. Knowing that $\mathcal{T} \in L(Z)$ we deduce that $\mathcal{T} \in L(Z, H)$ because

$$\|\mathcal{T}z\|_H \leq c \|\mathcal{T}z\|_Z \leq c \|\mathcal{T}\|_{L(Z)} \|z\|_Z \ \forall z \in Z.$$

Since $Z \subseteq H$ dense there is a unique operator $\mathcal{T}_H \in L(H)$ with $\mathcal{T}_H|_Z = \mathcal{T}$, cf. [Wer05, Satz II.1.5].

(ii) $\mathcal{T}^{-1}$ is continuous since $\mathcal{T}$ is continuous, cf. [Wer05, Korollar IV.3.4]. Using (i) we have the unique extension operators $\mathcal{T}_H \in L(H)$ resp. $(\mathcal{T}^{-1})_H \in L(H)$ of $\mathcal{T}$ resp. $\mathcal{T}^{-1}$. We show $(\mathcal{T}^{-1})_H \mathcal{T}_H v = v$ for all $v \in H$. If $v \in Z$ there is no need for any further proof. If $v \in H \setminus Z$ there is a sequence $(v_n) \subseteq Z$ with $v_n \to v$ in H as $n \to \infty$. Since $(\mathcal{T}^{-1})_H \mathcal{T}_H v_n = v_n$ we have

$$\left\| (\mathcal{T}^{-1})_H \mathcal{T}_H v - v \right\|_H \leq \left\| (\mathcal{T}^{-1})_H \mathcal{T}_H (v - v_n) \right\|_H + \left\| v_n - v \right\|_H$$
$$\leq \left(\left\| (\mathcal{T}^{-1})_H \right\|_{L(H)} \left\| \mathcal{T}_H \right\|_{L(H)} + 1 \right) \left\| v_n - v \right\|_H \to 0 \text{ as } n \to \infty.$$

Hence $(\mathcal{T}^{-1})_H \mathcal{T}_H v = v$ and analogously it can be shown that $\mathcal{T}_H (\mathcal{T}^{-1})_H v = v$ for all $v \in H$.

(iii) Making use of (i) and (ii) we obtain $c > 0$ such that

$$\|v\|_H = \left\| \mathcal{T}_H^{-1} \mathcal{T}_H v \right\|_H \leq \left\| \mathcal{T}_H^{-1} \right\|_{L(H)} \left\| \mathcal{S}^{*H} \mathcal{S} v \right\|_H \overset{\mathcal{S}^{*H} \in L(H)}{\leq} c \left\| \mathcal{S} v \right\|_H \quad \forall v \in H.$$

(iv) Let $v \in H$. Then we have

$$(v | \mathcal{T}_H v)_H = \left\| \mathcal{S} v \right\|_H^2 \geq c \left\| v \right\|_H^2$$

for a $c > 0$ using (iii) and that $\mathcal{T}_H = \mathcal{S}^{*H} \mathcal{S}$. $\qquad\square$

To illustrate the properties of $\mathcal{T}$ we provide a short example.

Example 4.9.
Consider the evolution triple $Z \subseteq H \subseteq Z^*$ with $Z = (H_0^1(\Omega))^2$ and $H = (L_2(\Omega))^2$ with $\Omega = (a, b) \subseteq \mathbb{R}$. Then $Z^* = (H^{-1}(\Omega))^2$. Let $a_{ij} \in C^1(\Omega) \cap C(\overline{\Omega})$ for $i, j = 1, 2$. Then we define $\mathcal{T}$ as follows.

$$\mathcal{T} : Z \to Z, \quad \mathcal{T} : \begin{pmatrix} z_1 \\ z_2 \end{pmatrix} \mapsto \begin{pmatrix} a_{11} z_1 + a_{12} z_2 \\ a_{21} z_1 + a_{22} z_2 \end{pmatrix}$$

The operator $\mathcal{T}$ is well-defined because $a_{ij} z_k \in H_0^1(\Omega)$ for $i, j, k = 1, 2$ and $\mathcal{T}$ is linear and continuous. If we require the matrix $A(x) := (a_{ij}(x))_{i,j=1,2}$ to be symmetric positive definite for all $x \in \overline{\Omega}$, then $\mathcal{T}$ fulfills Assumption 4.7. This can be seen as follows. We fix $x \in \Omega$ and omit it as an argument. We have $a_{12} = a_{21}$, the matrix A is invertible and

$$A^{-1} = \frac{1}{a_{11} a_{22} - a_{12}^2} \begin{pmatrix} a_{22} & -a_{12} \\ -a_{12} & a_{11} \end{pmatrix}.$$

All entries of A^{-1} are continuously differentiable. So by the means of A^{-1} the operator $\mathcal{T}$ is invertible and $\mathcal{T}_H \in L(H)$. $\mathcal{T}_H$ is self-adjoint because for $v, w \in H$ we have

$$(v | \mathcal{T}_H w)_H = (v_1 | a_{11} w_1 + a_{12} w_2)_{L_2(\Omega)} + (v_2 | a_{12} w_1 + a_{22} w_2)_{L_2(\Omega)}$$
$$= (a_{11} v_1 + a_{12} v_2 | w_1)_{L_2(\Omega)} + (a_{12} v_1 + a_{22} v_2 | w_2)_{L_2(\Omega)}$$
$$= (\mathcal{T}_H v | w)_H.$$

Since A is symmetric positive definite we can use the Cholesky decomposition to obtain a lower triangular matrix L such that $A = LL^\top$. With $L = (l_{ij})_{i,j=1,2}$ we set

$$\mathcal{S} : H \to H, \quad \mathcal{S} : \begin{pmatrix} w_1 \\ w_2 \end{pmatrix} \mapsto \begin{pmatrix} l_{11}w_1 + l_{21}w_2 \\ l_{22}w_2 \end{pmatrix}.$$

It is $\mathcal{S} \in L(H)$ because $l_{ij} \in C^1(\Omega) \cap C(\overline{\Omega})$ (can be verified by simple calculation) and

$$\mathcal{S}^{*_H} : H \to H, \quad \mathcal{S}^{*_H} : \begin{pmatrix} w_1 \\ w_2 \end{pmatrix} \mapsto \begin{pmatrix} l_{11}w_1 \\ l_{21}w_1 + l_{22}w_2 \end{pmatrix}.$$

It is then easy to verify that $\mathcal{T}_H = \mathcal{S}^{*_H}\mathcal{S}$.

The conditions on the operator $\mathcal{T}$ are such that it behaves well with the integration by parts formula. This will be elaborated in the following lemmata.

Lemma 4.10.
Let Assumption 4.1 be fulfilled and $\mathcal{T}$ as in Assumption 4.7. Let be $w \in W_2^1$ and define

$$\mathcal{T}w : \mathcal{I} \to Z, \quad (\mathcal{T}w)(t) := \mathcal{T}w(t) \quad t \in \mathcal{I}.$$

Then $\mathcal{T}w \in W_2^1$. Furthermore

$$\langle w'(t), \mathcal{T}w(t) \rangle_Z = \langle [\mathcal{T}w]'(t), w(t) \rangle_Z \quad f.a.a. \ t \in \mathcal{I}.$$

PROOF:
It has to be proven that

(i) $\mathcal{T}w \in L_2(\mathcal{I}, Z)$

(ii) $[\mathcal{T}w]' \in L_2(\mathcal{I}, Z^*)$, i.e. there exists $f \in L_2(\mathcal{I}, Z^*)$ such that

$$\int_{\mathcal{I}} (\mathcal{T}w(t)|z)_H \varphi'(t)\mathrm{d}t = - \int_{\mathcal{I}} \langle f(t), z \rangle_Z \varphi(t)\mathrm{d}t \quad \forall z \in Z, \ \varphi \in C_0^\infty(\mathcal{I})$$

(i) Since w is integrable, so is $\mathcal{T}w$, cf. [Emm04, Satz 7.1.15, (iii)]. So $\mathcal{T}w$ is especially measurable. We then have

$$\|\mathcal{T}w\|_{L_2(\mathcal{I},Z)}^2 \leq \int_{\mathcal{I}} \|\mathcal{T}\|_{L(Z)}^2 \|w(t)\|_Z^2 \, \mathrm{d}t = \|\mathcal{T}\|_{L(Z)}^2 \|w\|_{L_2(\mathcal{I},Z)}^2 < \infty$$

and so $\mathcal{T}w \in L_2(\mathcal{I}, Z)$.
(ii) Let $\mathcal{T}^* \in L(Z^*)$ be the dual operator of $\mathcal{T}$. We set $f(t) := \mathcal{T}^*w'(t)$. The same argument as above gives the measurability of f and we derive

$$\|f\|_{L_2(\mathcal{I},Z^*)}^2 = \int_{\mathcal{I}} \|\mathcal{T}^*w'(t)\|_{Z^*}^2 \, \mathrm{d}t \leq \|\mathcal{T}^*\|_{L(Z^*)}^2 \|w'\|_{L_2(\mathcal{I},Z^*)}^2 < \infty.$$

Using the properties of $\mathcal{T}$ and $\mathcal{T}_H$ we verify for all $z \in Z$ and $\varphi \in C_0^\infty(\mathcal{I})$ that

$$(w(t)\,|\,\mathcal{T}z)_H = (w(t)\,|\,\mathcal{T}_H z)_H = (\mathcal{T}_H w(t)\,|\,z)_H = (\mathcal{T}w(t)\,|\,z)_H$$

and thus

$$
\begin{aligned}
-\int_\mathcal{I} \langle f(t), z \rangle_Z \varphi(t)\mathrm{dt} &= -\int_\mathcal{I} \langle \mathcal{T}^* w'(t), z \rangle_Z \varphi(t)\mathrm{dt} \\
&= -\int_\mathcal{I} \langle w'(t), \mathcal{T}z \rangle_Z \varphi(t)\mathrm{dt} \\
&= \int_\mathcal{I} (w(t)\,|\,\mathcal{T}z)_H \varphi'(t)\mathrm{dt} \\
&= \int_\mathcal{I} (\mathcal{T}w(t)\,|\,z)_H \varphi'(t)\mathrm{dt}
\end{aligned}
$$

which implies $\mathcal{T}^* w' = f = [\mathcal{T}w]'$. This proves the lemma. $\qquad\square$

Lemma 4.11.
Let Assumption 4.1 be fulfilled and let be $w \in W_2^1$ and $\alpha \in C^\infty(\mathcal{I})$. Then $\alpha w \in W_2^1$ where $(\alpha w)(t) := \alpha(t)w(t)$ and we have the product rule

$$(\alpha w)'(t) = \alpha'(t)w(t) + \alpha(t)w'(t).$$

With $\alpha'(t)w(t)$ we mean $\alpha'(t)j(w(t))$ where $\alpha'(t)$ is the derivative of $\alpha(t)$ in $C^\infty(\mathcal{I})$ and $j \in L(H, Z^*)$ is the natural embedding as presented in Proposition A.6.

PROOF:
First we have that $\alpha w \in L_2(\mathcal{I}, Z)$ because it is measurable and

$$\|\alpha w\|_{L_2(\mathcal{I},Z)}^2 \leq \max_{t \in \mathcal{I}} |\alpha(t)|^2 \|w\|_{L_2(\mathcal{I},Z)}^2 < \infty.$$

Moreover we define $f : \mathcal{I} \to Z^*$ as

$$f(t) := \alpha'(t)j(w(t)) + \alpha(t)w'(t).$$

The measurability is clear and $f \in L_2(\mathcal{I}, Z^*)$ because

$$
\begin{aligned}
\|f\|_{L_2(\mathcal{I},Z^*)}^2 &\leq \widetilde{c}\,\|\alpha'\|_\infty^2 \|j\|_{L(H,Z^*)}^2 \int_\mathcal{I} \|w(t)\|_H^2 \,\mathrm{dt} + \|\alpha\|_\infty^2 \|w'\|_{L_2(\mathcal{I},Z^*)}^2 \\
&\leq C\,\|w\|_{W_2^1}^2 < \infty.
\end{aligned}
$$

for constants $\widetilde{c}$, $C > 0$. We used that the embedding $Z \subseteq H$ is continuous. Furthermore for all $z \in Z$ and $\varphi \in C_0^\infty(\mathcal{I})$ we obtain

$$
\begin{aligned}
\int_{\mathcal{I}} \langle f(t), z \rangle_Z \varphi(t) \mathrm{d}t \;&=\; \int_{\mathcal{I}} \langle \alpha(t) w'(t), z \rangle_Z \varphi(t) \mathrm{d}t + \int_{\mathcal{I}} \langle \alpha'(t) j(w(t)), z \rangle_Z \varphi(t) \mathrm{d}t \\[4pt]
&\overset{\alpha\varphi \in C_0^\infty}{=}\; -\int_{\mathcal{I}} (w(t) \,|\, z)_H (\alpha\varphi)'(t) \mathrm{d}t + \int_{\mathcal{I}} (\alpha'(t) w(t) \,|\, z)_H \varphi(t) \mathrm{d}t \\[4pt]
&=\; -\int_{\mathcal{I}} (w(t) \,|\, z)_H \left[(\alpha\varphi)'(t) - \alpha'(t)\varphi(t) \right] \mathrm{d}t \\[4pt]
&=\; -\int_{\mathcal{I}} (w(t) \,|\, z)_H \alpha(t) \varphi'(t) \mathrm{d}t \\[4pt]
&=\; -\int_{\mathcal{I}} (\alpha(t) w(t) \,|\, z)_H \varphi'(t) \mathrm{d}t
\end{aligned}
$$

using the product rule for continuously differentiable real valued functions. Hence we have $f = (\alpha w)'$ and the desired product rule follows immediately. $\qquad\square$

Lemma 4.12.
Let Assumption 4.1 be fulfilled and $\mathcal{T}$ be defined as in Assumption 4.7. Let $w \in W_2^1$. Then

$$
\frac{\mathrm{d}}{\mathrm{d}t} \|\mathcal{S}w(t)\|_H^2 = 2 \langle w'(t), \mathcal{T}w(t) \rangle_Z
$$

where $\dfrac{\mathrm{d}}{\mathrm{d}t}$ denotes the weak derivative of a real valued function, i.e.

$$
\int_{\mathcal{I}} \frac{\mathrm{d}}{\mathrm{d}t} \|\mathcal{S}w(t)\|_H^2 \, \varphi(t) \mathrm{d}t = -\int_{\mathcal{I}} \|\mathcal{S}w(t)\|_H^2 \, \varphi'(t) \mathrm{d}t \quad \forall \varphi \in C_0^\infty(\mathcal{I})
$$

PROOF:
It is $w \in L_2(\mathcal{I}, Z) \subseteq L_2(\mathcal{I}, H)$ and $w' \in L_2(\mathcal{I}, Z^*)$. Remembering the comment in the proof of Lemma 4.10 (i) we see that $\mathcal{S}w$ is measurable. Then also the function

$$
t \mapsto \|\mathcal{S}w(t)\|_H^2
$$

is measurable, even integrable because

$$
\begin{aligned}
\left\| \left(\|\mathcal{S}w(\cdot)\|_H^2 \right) \right\|_{L_1(\mathcal{I})} \;&\leq\; \|\mathcal{S}\|_{L(H)}^2 \int_{\mathcal{I}} \|w(t)\|_H^2 \, \mathrm{d}t \\[4pt]
&\overset{w(t) \in Z}{\leq}\; c \, \|\mathcal{S}\|_{L(H)}^2 \int_{\mathcal{I}} \|w(t)\|_Z^2 \, \mathrm{d}t \\[4pt]
&\leq\; c \, \|\mathcal{S}\|_{L(H)}^2 \, \|w\|_{L_2(\mathcal{I},Z)}^2 < \infty.
\end{aligned}
$$

The function

$$
t \mapsto \langle w'(t), \mathcal{T}w(t) \rangle_Z
$$

is also integrable, i.e. in $L_1(\mathcal{I})$. This is a direct consequence of the Hölder inequality (Proposition A.7):

$$
\begin{aligned}
\|\langle w'(\cdot), \mathcal{T}w(\cdot)\rangle_Z\|_{L_1(\mathcal{I})} &= \int_{\mathcal{I}} |\langle w'(t), \mathcal{T}w(t)\rangle_Z| \, \mathrm{d}t \\
&\leq \|w'\|_{L_2(\mathcal{I},Z^*)} \|\mathcal{T}w\|_{L_2(\mathcal{I},Z)} < \infty
\end{aligned}
$$

because $\mathcal{T}w \in L_2(\mathcal{I}, Z)$, cf. Lemma 4.10. Let be $\varphi \in C_0^\infty(\mathcal{I})$ and set $v(t) := \varphi(t)\mathcal{T}w(t)$. We can use Lemma 4.10 to get $\mathcal{T}w \in W_2^1$ because $w \in W_2^1$ and $\mathcal{T}$ satisfies the relevant properties. Furthermore with Lemma 4.11 we obtain $v \in W_2^1$. We now use the integration by parts formula from Proposition A.10 for $s = t_0$, $t = T$. Applying the product rule from Lemma 4.11 we get

$$
\begin{aligned}
0 &= \int_{\mathcal{I}} \langle w'(t), v(t)\rangle_Z + \langle v'(t), w(t)\rangle_Z \mathrm{d}t \\
&= \int_{\mathcal{I}} \langle w'(t), \mathcal{T}w(t)\rangle_Z \varphi(t) + \langle (\mathcal{T}w)'(t), w(t)\rangle_Z \varphi(t) + (\mathcal{T}w(t)\,|\,w(t))_H \varphi'(t) \mathrm{d}t \\
&= 2 \int_{\mathcal{I}} \langle w'(t), \mathcal{T}w(t)\rangle_Z \varphi(t)\mathrm{d}t + \int_{\mathcal{I}} (\mathcal{S}w(t)\,|\,\mathcal{S}w(t))_H \varphi'(t)\mathrm{d}t.
\end{aligned}
$$

The last line follows from Lemma 4.10 and we directly deduce that $2\langle w'(t), \mathcal{T}w(t)\rangle_Z$ is the weak derivative of $\|\mathcal{S}w(t)\|_H^2$. $\qquad\square$

Corollary 4.13.
Let Assumption 4.1 be fulfilled and $\mathcal{T}$ be defined as in Assumption 4.7. Let be $u \in W_{2,\mathcal{D}}^1$ and s,t with $t_0 \leq s \leq t \leq T$. Then

$$
\|\mathcal{S}\mathcal{D}u(t)\|_H^2 - \|\mathcal{S}\mathcal{D}u(s)\|_H^2 = 2\int_s^t \langle [\mathcal{D}u(\tau)]', \mathcal{A}u(\tau)\rangle_Z \mathrm{d}\tau
$$

PROOF:
Let $u \in W_{2,\mathcal{D}}^1$ and therefore $\mathcal{D}u \in W_2^1$. An application of Lemma 4.12 for $w = \mathcal{D}u$ gives the desired result after integrating over $[s,t]$. $\qquad\square$

The operator family $\mathcal{B}(t)$

In this section we introduce conditions for the operator family $\mathcal{B}(t)$ in order to obtain a solvability result. In the standard theory of nonlinear evolution equations with monotone operators the same or similar conditions are required, see [Rou05] or [Zei90b].

Assumption 4.14 (Operator family $\mathcal{B}(t)$).
The operator family

$$
\mathcal{B}(t) : V \to V^*, \; t \in \mathcal{I}
$$

fulfills the following properties.

(i) Strong monotonicity. There exists $\mu > 0$ such that

$$\langle \mathcal{B}(t)(u) - \mathcal{B}(t)(v), u - v \rangle_V \geq \mu \|u - v\|_V^2 \quad \forall u, v \in V, \ t \in \mathcal{I}.$$

(ii) Hemicontinuity. The map

$$s \mapsto \langle \mathcal{B}(t)(u + sv), w \rangle_V$$

is continuous on $[0, 1]$ for all $u, v, w \in V$, $t \in \mathcal{I}$.

(iii) Growth condition. There exists a non-negative function $g \in L_2(\mathcal{I})$ and a constant $c > 0$ such that

$$\|\mathcal{B}(t)(v)\|_{V^*} \leq g(t) + c \|v\|_V \quad \forall v \in V, \ t \in \mathcal{I}.$$

(iv) Measurability. The map $t \mapsto \mathcal{B}(t)$ is weakly measurable, i.e. the function

$$t \mapsto \langle \mathcal{B}(t)(u), v \rangle_V$$

is measurable on $\mathcal{I}$ for all $u, v \in V$.

We set $X := L_2(\mathcal{I}, V)$ and hence $X^* = L_2(\mathcal{I}, V^*)$. We define

$$\mathcal{B} : X \to X^*, \quad (\mathcal{B}u)(t) := \mathcal{B}(t)(u(t)) \quad \forall u \in X, \ t \in \mathcal{I} \tag{4.5}$$

and have

$$\langle \mathcal{B}u, v \rangle_X = \int_{\mathcal{I}} \langle (\mathcal{B}(t)u(t)), v(t) \rangle_V \mathrm{d}t \quad \forall u, v \in X. \tag{4.6}$$

The operator $\mathcal{B} : X \to X^*$ inherits the properties of the operator family $\mathcal{B}(t)$ as it is shown in the next lemma.

Lemma 4.15.
Let the operator family $\mathcal{B}(t)$ fulfill Assumption 4.14. Then the operator $\mathcal{B}$ as defined in (4.5) satisfies the following.

(i) Strong monotonicity. There exists $\widetilde{\mu} > 0$ such that

$$\langle \mathcal{B}u - \mathcal{B}v, u - v \rangle_X \geq \widetilde{\mu} \|u - v\|_X^2 \quad \forall u, v \in X.$$

(ii) Hemicontinuity. The map

$$s \mapsto \langle \mathcal{B}(u + sv), w \rangle_X$$

is continuous on $[0, 1]$ for all $u, v, w \in X$.

(iii) *Growth condition.* *There exists a non-negative function* $g \in L_2(\mathcal{I})$ *and a constant* $c > 0$ *such that*

$$\|\mathcal{B}v\|_{X^*} \leq c(\|g\|_{L_2(\mathcal{I})} + \|v\|_X) \quad \forall v \in X.$$

(iv) *Measurability.* *For all* $u \in X$, $v \in V$ *the real function*

$$t \mapsto \langle(\mathcal{B}u)(t), v\rangle_V$$

is measurable on $\mathcal{I}$.

PROOF:
The measurability was proven in [Zei90b, Lemma 30.2 and Problem 30.1]. Having checked the measurability, (ii) and (iii) can be proven as in [Zei90b, section 30.3b]. It remains to show (i). Let $u, v \in X$. The Hölder inequality A.7 tells us that the real function

$$t \mapsto \langle\mathcal{B}(t)(u(t)), v(t)\rangle_V$$

is integrable over $\mathcal{I}$. Since $\mathcal{B}(t) : V \to V^*$ is strongly monotone with a uniform $\mu > 0$ we have

$$
\begin{aligned}
\langle\mathcal{B}u - \mathcal{B}v, u - v\rangle_X &= \int_{\mathcal{I}} \langle\mathcal{B}(t)(u(t)) - \mathcal{B}(t)(v(t)), u(t) - v(t)\rangle_V \mathrm{d}t \\
&\geq \int_{\mathcal{I}} \mu \|u(t) - v(t)\|_V^2 \, \mathrm{d}t \\
&\geq \mu \|u - v\|_X^2
\end{aligned}
$$

with $\widetilde{\mu} = \mu$. $\qquad\square$

Lemma 4.16 (Uniqueness of solutions).
Let Assumptions 4.1, 4.3, 4.4, 4.7 and 4.14 be fulfilled. Then a solution $u \in W_{2,\mathcal{D}}^1$ *to* (4.4) *is unique.*

PROOF:
Let $u_1, u_2 \in W_{2,\mathcal{D}}^1$ be two solutions of (4.4) in $W_{2,\mathcal{D}}^1$. Then we obtain using the linearity of $\mathcal{D}$ and of the generalized derivative for all $v \in V$ and f.a.a. $t \in \mathcal{I}$:

$$\langle[\mathcal{D}(u_1(t) - u_2(t))]', \mathcal{A}v\rangle_Z + \langle\mathcal{B}(t)(u_1(t)) - \mathcal{B}(t)(u_2(t)), v\rangle_V = 0.$$

We insert $u_1(t) - u_2(t) \in V$ for v and integrate over $\mathcal{I}$. Since $u_1 - u_2 \in W_{2,\mathcal{D}}^1$ the integration by parts formula 4.13 leads to

$$
\begin{aligned}
0 &= \|\mathcal{S}\mathcal{D}u_1(T) - \mathcal{S}\mathcal{D}u_2(T)\|_H^2 - \|\mathcal{S}\mathcal{D}u_1(t_0) - \mathcal{S}\mathcal{D}u_2(t_0)\|_H^2 \\
&\quad + 2\int_{\mathcal{I}} \langle\mathcal{B}(t)(u_1(t)) - \mathcal{B}(t)(u_2(t)), u_1(t) - u_2(t)\rangle_V \mathrm{d}t \\
&\geq \|\mathcal{S}\mathcal{D}(u_1(T) - u_2(T))\|_H^2 + 2\mu \int_{\mathcal{I}} \|u_1(t) - u_2(t)\|_V^2 \, \mathrm{d}t.
\end{aligned}
$$

Here we used the strong monotonicity of $\mathcal{B}(t)$ and that $\mathcal{D}u_1(t_0) = z_0 = \mathcal{D}u_2(t_0)$. Hence

$$0 \leq \int_{\mathcal{I}} \|u_1(t) - u_2(t)\|_V^2 \, dt \leq 0$$

and consequently $u_1(t) = u_2(t)$ f.a.a. $t \in \mathcal{I}$. $\qquad\qquad\qquad\qquad\qquad\qquad\square$

4.2. The Galerkin equations

In this section we derive the Galerkin equations of problem (4.4) and prove unique solvability. Without mentioning them explicitly we will require Assumptions 4.1, 4.3, 4.4, 4.7 and 4.14 in the following. Furthermore we make the following assumption concerning the Galerkin scheme.

Assumption 4.17 (Basis).
Let $\{w_1, w_2, \dots\} \subseteq V$ be a basis in the sense of Definition A.11 with the following properties:

(i) There exists a sequence $(z_{n0}) \subseteq Z \subseteq H$ with

$$z_{n0} \to z_0 \text{ in } H, \text{ i.e. } \|z_{n0} - z_0\|_H \to 0 \text{ as } n \to \infty$$

where

$$z_{n0} \in \operatorname{span}\{\mathcal{D}w_1, \dots \mathcal{D}w_n\} \quad \forall n \in \mathbb{N}.$$

(ii) The basis $\{w_1, w_2, \dots\}$ of V is chosen such that $w_i \in \operatorname{im} \mathcal{P}$ or $w_i \in \operatorname{im} \mathcal{Q}$ for all i with $\mathcal{P}$ and $\mathcal{Q}$ being the projection operators from Assumption 4.4. Let

$$I_{\mathcal{P}} := \{i|\ w_i \in \operatorname{im} \mathcal{P}\}, \quad I_{\mathcal{Q}} := \{i|\ w_i \in \operatorname{im} \mathcal{Q}\}$$

and for $n \in \mathbb{N}$ let

$$I_{n\mathcal{P}} := \{i|\ 1 \leq i \leq n, \ w_i \in \operatorname{im} \mathcal{P}\}, \quad n\mathcal{P} := |I_{n\mathcal{P}}|,$$
$$I_{n\mathcal{Q}} := \{i|\ 1 \leq i \leq n, \ w_i \in \operatorname{im} \mathcal{Q}\}, \quad n\mathcal{Q} := |I_{n\mathcal{Q}}|$$

then

$$I_{n\mathcal{P}} \to I_{\mathcal{P}} \quad \text{and} \quad I_{n\mathcal{Q}} \to I_{\mathcal{Q}} \quad \text{as } n \to \infty.$$

In [Tis04] it is required that there is a basis $\{z_1, z_2, \dots\} \subseteq Z$ and that for all $n \in \mathbb{N}$ there exists an $m_n \in \mathbb{N}$ such that

$$\{\mathcal{D}w_1, \dots, \mathcal{D}w_n\} \subseteq \operatorname{span}\{z_1, \dots, z_{m_n}\}.$$

This condition is not needed. Furthermore we remark that the assumption on the basis is satisfied if it is ensured that

$$w_i \in \operatorname{im} \mathcal{P} \quad \text{for odd } i \quad \text{and} \quad w_i \in \operatorname{im} \mathcal{Q} \quad \text{for even } i.$$

If one of these spaces is finite dimensional we stop with this numbering and proceed with normal numbering for the other space.

We fix $n \in \mathbb{N}$. In order to set up the Galerkin equations we consider the finite dimensional spaces

$$V_n := \operatorname{span} \{w_1, \ldots, w_n\}$$

and

$$H_n := \operatorname{span} \{\mathcal{D}w_1, \ldots, \mathcal{D}w_n\}.$$

We equip H_n with the scalarproduct of H. We note that $\dim H_n \leq n$ and that

$$H_n \subseteq Z \subseteq H \subseteq Z^*.$$

We consider (4.4a) and replace formally $u(t) \in V$ by $u_n(t) \in V_n$ and $v \in V$ by $w_i \in V_n$. Thus the coordinate free Galerkin equations are obtained.

$$\langle [\mathcal{D}u_n(t)]', \mathcal{A}w_i \rangle_Z + \langle \mathcal{B}(t)(u_n(t)), w_i \rangle_V = \langle r(t), w_i \rangle_V \qquad (4.7\text{a})$$
$$\mathcal{D}u_n(t_0) = z_{n0} \qquad (4.7\text{b})$$

for all $i = 1, \ldots n$ and f.a.a. $t \in \mathcal{I}$. The solution space of (4.7) will be

$$W_n := \left\{ u_n \in L_2(\mathcal{I}, V_n) \mid [\mathcal{D}u_n]' \in L_2(\mathcal{I}, H_n) \right\}.$$

Since the embedding $H \subseteq Z^*$ is continuous it is

$$\langle [\mathcal{D}u_n(t)]', \mathcal{A}w_i \rangle_Z = \frac{\mathrm{d}}{\mathrm{d}t} (\mathcal{D}u_n(t) \,|\, \mathcal{A}w_i)_H \qquad \text{f.a.a. } t \in \mathcal{I}$$

where $\frac{\mathrm{d}}{\mathrm{d}t}$ is the weak derivative of a real valued function. We will now prove a priori estimates for the solutions of the Galerkin equations. A simple but helpful tool is the following classical inequality. For all $\alpha > 0$ it holds that

$$2|xy| \leq \alpha^{-1}x^2 + \alpha y^2 \quad \forall x, y \in \mathbb{R}. \qquad (4.8)$$

A direct consequence is

$$(x + y)^2 \leq 2x^2 + 2y^2 \quad \forall x, y \in \mathbb{R}.$$

We will use these classical inequalities in the next lemma and later on.

Lemma 4.18 (A priori estimates).
Let Assumptions 4.1, 4.3, 4.4, 4.7, 4.14 and 4.17 be fulfilled. Let $u_n \in W_n$ be a solution
of (4.7). Then the following holds.

(i) $u_n \in W^1_{2,\mathcal{D}}$.

(ii) u_n is unique.

(iii) There exists a $C > 0$, independent of n, such that

$$\|u_n\|_{L_2(\mathcal{I},V)} \le C,$$
$$\|\mathcal{B}u_n\|_{L_2(\mathcal{I},V^*)} \le C,$$
$$\max_{t \in \mathcal{I}} \|\mathcal{D}u_n(t)\|_H \le C.$$

PROOF:
(i) We have $H_n \subseteq Z \subseteq H \subseteq Z^*$ and $\dim H_n < \infty$. There is a $c_1 > 0$ such that

$$\|z\|_Z \le c_1 \|z\|_H \quad \forall z \in H_n$$

because the norms $\|\cdot\|_Z$ and $\|\cdot\|_H$ are equivalent on H_n. Additionally we have for all
$x, z \in H_n$ that

$$\langle z, x \rangle_Z = (z \,|\, x)_H \le \|z\|_H \|x\|_H \le c_2 \|z\|_H \|x\|_Z$$

for a $c_2 > 0$ because $Z \subseteq H \subseteq Z^*$ is an evolution triple. So it is

$$\|z\|_{Z^*} \le c \|z\|_H \quad \forall z \in H_n$$

with a $c > 0$. Then it follows from $u_n \in W_n$ that $u_n \in L_2(\mathcal{I}, V)$ and $[\mathcal{D}u_n]' \in L_2(\mathcal{I}, Z^*)$
proving (i).
(ii) Because of (i) the integration by parts formula 4.13 can be applied to u_n and so (ii)
follows from Lemma 4.16 using the same arguments.
(iii) Let $u_n \in W_n \subseteq W^1_{2,\mathcal{D}}$ be a solution of (4.7). Then it is

$$\langle [\mathcal{D}u_n(t)]', \mathcal{A}u_n(t) \rangle_Z + \langle \mathcal{B}(t)(u_n(t)) - r(t), u_n(t) \rangle_V = 0 \quad \text{f.a.a. } t \in \mathcal{I}.$$

Integrating over $[t_0, t]$, $t \le T$ and applying the integration by parts formula 4.13 gives

$$
\begin{aligned}
0 &= 2 \int_{t_0}^{t} \langle [\mathcal{D}u_n(s)]', \mathcal{A}u_n(s) \rangle_Z + \langle \mathcal{B}(s)(u_n(s)) - r(s), u_n(s) \rangle_V \mathrm{d}s \\
&= \|\mathcal{S}\mathcal{D}u_n(t)\|_H^2 - \|\mathcal{S}\mathcal{D}u_n(t_0)\|_H^2 \\
&\quad + 2 \int_{t_0}^{t} \langle \mathcal{B}(s)(u_n(s)) - \mathcal{B}(s)(0) + \mathcal{B}(s)(0) - r(s), u_n(s) \rangle_V \mathrm{d}s \\
&\ge \|\mathcal{S}\mathcal{D}u_n(t)\|_H^2 - \|\mathcal{S}\mathcal{D}u_n(t_0)\|_H^2 \\
&\quad + 2\mu \int_{t_0}^{t} \|u_n(s)\|_V^2 \, \mathrm{d}s + 2 \int_{t_0}^{t} \langle \mathcal{B}(s)(0) - r(s), u_n(s) \rangle_V \mathrm{d}s
\end{aligned}
$$

In the last line we used the strong monotonicity of $\mathcal{B}(s)$. Rearranging terms gives

$$\|\mathcal{SD}u_n(t)\|_H^2 + 2\mu \int_{t_0}^t \|u_n(s)\|_V^2\,\mathrm{d}s \le \|\mathcal{SD}u_n(t_0)\|_H^2 + 2\int_{t_0}^t \langle r(s) - \mathcal{B}(t)(0), u_n(s)\rangle_V \mathrm{d}s$$

Applying the classical inequality (4.8) with $2\alpha = \mu$ gives

$$
\begin{aligned}
& 2\int_{t_0}^t \langle r(s) - \mathcal{B}(t)(0), u_n(s)\rangle_V \mathrm{d}s \\
\le\ & 2\int_{t_0}^t \|r(s)\|_{V^*}\|u_n(s)\|_V + 2\int_{t_0}^t \|\mathcal{B}(s)(0)\|_{V^*}\|u_n(s)\|_V\,\mathrm{d}s \\
\le\ & \alpha^{-1}\int_{t_0}^t \|r(s)\|_{V^*}^2\,\mathrm{d}s + \alpha^{-1}\int_{t_0}^t \|\mathcal{B}(s)(0)\|_{V^*}^2\,\mathrm{d}s + \mu\int_{t_0}^t \|u_n(s)\|_V^2\,\mathrm{d}s \\
\le\ & 2\mu^{-1}\|g\|_{L_2(\mathcal{I})}^2 + 2\mu^{-1}\|r\|_{L_2(\mathcal{I},V^*)}^2 + \mu\|u_n\|_{L_2(\mathcal{I},V)}^2\,.
\end{aligned}
$$

Here we used the growth condition for $\mathcal{B}(t)$. Inserting $t = T$ we obtain

$$\|\mathcal{SD}u_n(T)\|_H^2 + \mu\|u_n\|_{L_2(\mathcal{I},V)}^2 \le \|\mathcal{S}z_{0n}\|_H^2 + 2\mu^{-1}\|g\|_{L_2(\mathcal{I})}^2 + 2\mu^{-1}\|r\|_{L_2(\mathcal{I},V^*)}^2 \qquad (4.9)$$

Since $\|\mathcal{SD}u_n(T)\|_H^2 \ge 0$, $r \in L_2(\mathcal{I},V^*)$, $g \in L_2(\mathcal{I})$ and $z_{n0} \to z_0$ in H as $n \to \infty$ there is a $C_1 > 0$, independent of n, such that

$$\|u_n\|_{L_2(\mathcal{I},V)} \le C_1 \qquad (4.10)$$

Furthermore we see with the growth condition for $\mathcal{B}(t)$ and the Hölder inequality that

$$
\begin{aligned}
\int_{\mathcal{I}} \|\mathcal{B}(t)(u_n(t))\|_{V^*}^2\,\mathrm{d}t &\le \int_{\mathcal{I}} (g(t) + c\|u(t)\|_V)^2 \mathrm{d}t \\
&\le \int_{\mathcal{I}} 2g(t)^2 + 2c^2\|u_n(t)\|_V^2\,\mathrm{d}t \\
&\le 2\|g\|_{L_2(\mathcal{I})}^2 + 2c^2\|u_n\|_{L_2(\mathcal{I},V)}^2
\end{aligned}
$$

Using (4.10) there exists a $C_2 > 0$, independent of n, such that

$$\|\mathcal{B}u_n\|_{L_2(\mathcal{I},V^*)} \le C_2.$$

In analogy to the estimate (4.9) we arrive at

$$\|\mathcal{SD}u_n(t)\|_H^2 + \mu\int_{t_0}^t \|u(s)\|_V^2\,\mathrm{d}s \le \widetilde{C}$$

for a $\widetilde{C} > 0$, independent of n. We have

$$\int_{t_0}^t \|u(s)\|_V^2\,\mathrm{d}s \ge 0, \quad \|\mathcal{D}u(t)\|_H^2 \le c_1\|\mathcal{SD}u(t)\|_H^2$$

for a constant $c_1 > 0$, cf. Lemma 4.8 (iii). We conclude that there must be a $C_3 > 0$ such that

$$\max_{t \in \mathcal{I}} \|\mathcal{D}u(t)\|_H \leq C_3$$

Choosing $C = \max\{C_1, C_2, C_3\}$ gives the desired result and C does not depend on n. $\square$

For applying solvability results from the theory of ordinary differential equations we rewrite the coordinate free Galerkin equations (4.7) as follows. For fixed $n \in \mathbb{N}$ we represent

$$u_n(t) = \sum_{j=1}^{n} c_{jn}(t)w_j \in V_n$$

with coefficients $c_{jn}(t) \in \mathbb{R}$ for all $t \in \mathcal{I}$. We split

$$u_n(t) = \sum_{j \in I_{n\mathcal{P}}} c_{jn}(t)w_j + \sum_{j \in I_{n\mathcal{Q}}} c_{jn}(t)w_j \tag{4.11}$$

with $I_{n\mathcal{P}}$ and $I_{n\mathcal{Q}}$ as in Assumption 4.17. The index sets $I_{n\mathcal{P}}$ and $I_{n\mathcal{Q}}$ are disjoint and satisfy

$$I_{n\mathcal{P}} \cup I_{n\mathcal{Q}} = \{1, \ldots, n\}.$$

We set for $t \in \mathcal{I}$

$$x_j(t) := c_{jn}(t) \quad \forall j \in I_{n\mathcal{P}},$$
$$y_j(t) := c_{jn}(t) \quad \forall j \in I_{n\mathcal{Q}}$$

and

$$x(t) := (x_j(t))_{j \in I_{n\mathcal{P}}} \in \mathbb{R}^{n\mathcal{P}}, \quad y(t) := (y_j(t))_{j \in I_{n\mathcal{Q}}} \in \mathbb{R}^{n\mathcal{Q}}.$$

Inserting (4.11) into the Galerkin equations (4.7a) we obtain the system

$$\frac{\mathrm{d}}{\mathrm{d}t}\Big(\sum_{j \in I_{n\mathcal{P}}} x_j(t)\mathcal{D}w_j \,\big|\, \mathcal{T}\mathcal{D}w_i\Big)_H$$
$$+ \langle \mathcal{B}(t)(\sum_{j \in I_{n\mathcal{P}}} x_j(t)w_j + \sum_{j \in I_{n\mathcal{Q}}} y_j(t)w_j), w_i \rangle_V = \langle r(t), w_i \rangle \quad i \in I_{n\mathcal{P}}$$
$$\langle \mathcal{B}(t)(\sum_{j \in I_{n\mathcal{P}}} x_j(t)w_j + \sum_{j \in I_{n\mathcal{Q}}} y_j(t)w_j), w_i \rangle_V = \langle r(t), w_i \rangle \quad i \in I_{n\mathcal{Q}}$$

Here we used the linearity of the weak derivative and that

$$\mathcal{D}w_i = \mathcal{D}\mathcal{Q}w_i = 0 \quad \forall i \in I_{n\mathcal{Q}}.$$

The initial condition (4.7b) reads

$$\mathcal{D}u_n(t_0) = \sum_{j\in I_{n\mathcal{P}}} x_j(t_0)\mathcal{D}w_j = z_{0n} = \sum_{j\in I_{n\mathcal{P}}} x_{0j}\mathcal{D}w_j \tag{4.12}$$

with coefficients $x_{0j} \in \mathbb{R}$, $j \in I_{n\mathcal{P}}$ and $x_0 := (x_{0j})_{j\in I_{n\mathcal{P}}} \in \mathbb{R}^{n\mathcal{P}}$. Note that (4.12) is equivalent to $x(t_0) = x_0$ because the $\mathcal{D}w_j$, $j \in I_{n\mathcal{P}}$ are linearly independent.
Rewriting in matrix notation gives the system

$$
\begin{align}
Gx'(t) + b_{\mathcal{P}}(x(t), y(t), t) &= r_{\mathcal{P}}(t), \tag{4.13a}\\
b_{\mathcal{Q}}(x(t), y(t), t) &= r_{\mathcal{Q}}(t), \tag{4.13b}\\
x(t_0) &= x_0 \tag{4.13c}
\end{align}
$$

f.a.a. $t \in \mathcal{I}$ with

$$G := ((\mathcal{D}w_j \,|\, \mathcal{T}\mathcal{D}w_i)_H)_{i,j\in I_{n\mathcal{P}}} \in \mathbb{R}^{n\mathcal{P}\times n\mathcal{P}},$$

$$b_{\mathcal{P}}(x(t), y(t), t) := \left(\langle \mathcal{B}(t)(\sum_{j\in I_{n\mathcal{P}}} x_j(t)w_j + \sum_{j\in I_{n\mathcal{Q}}} y_j(t)w_j), w_i \rangle_V \right)_{i\in I_{n\mathcal{P}}} \in \mathbb{R}^{n\mathcal{P}},$$

$$b_{\mathcal{Q}}(x(t), y(t), t) := \left(\langle \mathcal{B}(t)(\sum_{j\in I_{n\mathcal{P}}} x_j(t)w_j + \sum_{j\in I_{n\mathcal{Q}}} y_j(t)w_j), w_i \rangle_V \right)_{i\in I_{n\mathcal{Q}}} \in \mathbb{R}^{n\mathcal{Q}},$$

$$r_{\mathcal{P}}(t) := (\langle r(t), w_i \rangle)_{i\in I_{n\mathcal{P}}} \in \mathbb{R}^{n\mathcal{P}},$$

$$r_{\mathcal{Q}}(t) := (\langle r(t), w_i \rangle)_{i\in I_{n\mathcal{Q}}} \in \mathbb{R}^{n\mathcal{Q}},$$

$$x_0 := (x_{0j})_{i\in I_{n\mathcal{P}}} \in \mathbb{R}^{n\mathcal{P}}.$$

We remark that all these terms depend on the Galerkin step n. It is shown in the next lemma that the Galerkin system (4.13) inherits important properties from the abstract system (4.4).

Lemma 4.19 (Properties of (4.13)).
Let Assumptions 4.1, 4.3, 4.4, 4.7, 4.14 and 4.17 be fulfilled. Then the following holds.

(i) G is positive definite.

(ii) $b_{\mathcal{Q}}$ is strongly monotone w.r.t. y.

(iii) $b_{\mathcal{P}}$, $b_{\mathcal{Q}}$ are continuous w.r.t. (x, y).

(iv) $b_\mathcal{P}$, $b_\mathcal{Q}$ *are measurable w.r.t. t.*

(v) $r_\mathcal{P} \in L_2(\mathcal{I}, \mathbb{R}^{n\mathcal{P}})$, $r_\mathcal{Q} \in L_2(\mathcal{I}, \mathbb{R}^{n\mathcal{Q}})$.

PROOF:
We define the following norms on $\mathbb{R}^{n\mathcal{P}}$ resp. $\mathbb{R}^{n\mathcal{Q}}$:

$$\|x\|_{n\mathcal{P}} := \left\| \sum_{j \in I_{n\mathcal{P}}} x_j w_j \right\|_V , \quad \|y\|_{n\mathcal{Q}} := \left\| \sum_{j \in I_{n\mathcal{Q}}} y_j w_j \right\|_V .$$

The norm properties are obviously fulfilled because the w_j are linearly independent.
(i) Let $x \in \mathbb{R}^{n\mathcal{P}}$. Then

$$x^T G x = \sum_{i,j \in I_{n\mathcal{P}}} x_i x_j (\mathcal{D} w_j | \mathcal{T} \mathcal{D} w_i)_H = (\sum_{j \in I_{n\mathcal{P}}} x_j \mathcal{D} w_j | \mathcal{T}_H (\sum_{i \in I_{n\mathcal{P}}} x_j \mathcal{D} w_j))_H \geq 0$$

because of the properties of $\mathcal{T}_H$, cf. Lemma 4.8 (iv). It is $x^T G x = 0$ if and only if $\sum_{j \in I_{n\mathcal{P}}} x_j \mathcal{D} w_j = 0$ since $\mathcal{T}_H$ is strongly monotone. This is also equivalent to

$$\sum_{j \in I_{n\mathcal{P}}} x_j w_j \in \ker \mathcal{D} = \operatorname{im} \mathcal{Q}.$$

We also have $\sum_{j \in I_{n\mathcal{P}}} x_j w_j \in \operatorname{im} \mathcal{P}$ and so $x_j = 0$ follows for all $j \in I_{n\mathcal{P}}$ because the w_j are linearly independent. Thus $x^T G x = 0$ if and only if $x = 0$, proving (i).
(ii) Let $x \in \mathbb{R}^{n\mathcal{P}}$, $t \in \mathcal{I}$ be fixed and let $y, \bar{y} \in \mathbb{R}^{n\mathcal{Q}}$. Making a zero addition and using that $\mathcal{B}(t)$ is strongly monotone we obtain

$$
\begin{aligned}
&(b_\mathcal{Q}(x, y, t) - b_\mathcal{Q}(x, \bar{y}, t) | y - \bar{y}) \\
= \ &\langle \mathcal{B}(t)(\sum_{j \in I_{n\mathcal{P}}} x_j w_j + \sum_{j \in I_{n\mathcal{Q}}} y_j w_j) - \mathcal{B}(t)(\sum_{j \in I_{n\mathcal{P}}} x_j w_j + \sum_{j \in I_{n\mathcal{Q}}} \bar{y}_j w_j), \sum_{j \in I_{n\mathcal{Q}}} (y_j - \bar{y}_j) w_j \rangle_V \\
\geq \ &\mu \|y - \bar{y}\|_{n\mathcal{Q}}^2 \geq \tilde{\mu} \|y - \bar{y}\|^2
\end{aligned}
$$

for a $\tilde{\mu} > 0$ because all norms on $\mathbb{R}^{n\mathcal{Q}}$ are equivalent. $\tilde{\mu}$ is independent of x and t.
(iii) We fix $t \in \mathcal{I}$. Let $((x_m, y_m)) \subseteq \mathbb{R}^n$ be a sequence with $(x_m, y_m) \to (x, y) \in \mathbb{R}^n$ as $m \to \infty$. We remark that $\mathcal{B}(t) : V \to V^*$ is monotone and hemicontinuous on the reflexive Banach space. Hence it is also demicontinuous, i.e.

$$v_m \to v \text{ in } V \quad \Rightarrow \quad \mathcal{B}(t)(v_m) \rightharpoonup \mathcal{B}(t)(v) \text{ in } V^* \quad \text{as } m \to \infty,$$

cf. [Zei90b, Proposition 26.4]. We set

$$v_m := \sum_{j \in I_{n\mathcal{P}}} x_{mj} w_j + \sum_{j \in I_{n\mathcal{Q}}} y_{mj} w_j, \quad v := \sum_{j \in I_{n\mathcal{P}}} x_j w_j + \sum_{j \in I_{n\mathcal{Q}}} y_j w_j$$

and obtain

$$\|v_m - v\|_V \le c_1 \left(\|x_m - x\|_{n\mathcal{P}} + \|y_m - y\|_{n\mathcal{Q}} \right)$$
$$\le c_2 \left(\|(x_m, y_m) - (x, y)\|_\infty \right) \to 0 \text{ as } m \to \infty$$

for constants $c_1, c_2 > 0$. Thus using the demicontinuity of $\mathcal{B}(t)$ we have for all $i \in I_{n\mathcal{P}}$:

$$\langle \mathcal{B}(t)(v_m), w_i \rangle_V \to \langle \mathcal{B}(t)(v), w_i \rangle_V \text{ as } m \to \infty.$$

This yields

$$\|b_\mathcal{P}(x_m, y_m, t) - b_\mathcal{P}(x, y, t)\|_\infty \to 0 \text{ as } m \to \infty.$$

The continuity of $b_\mathcal{Q}$ w.r.t. (x, y) follows analogously.
(iv) $b_\mathcal{P}$ and $b_\mathcal{Q}$ are measurable w.r.t. t because

$$t \mapsto \langle \mathcal{B}(t)(u), v \rangle_V$$

is measurable on $\mathcal{I}$ for all $u, v \in V$, cf. Assumption 4.14 (iv).
(v) This is a direct consequence of the fact that $r \in L_2(\mathcal{I}, V^*)$. $\qquad \square$

Lemma 4.20.
Let Assumptions 4.1, 4.3, 4.4, 4.7, 4.14 and 4.17 be fulfilled. Let (x, y) be a solution to (4.13) with $x \in C(\mathcal{I}, \mathbb{R}^{n\mathcal{P}})$, $y \in L_2(\mathcal{I}, \mathbb{R}^{n\mathcal{Q}})$ and the derivative $x'(t)$ exists f.a.a. $t \in \mathcal{I}$. Then

$$u_n(t) := \sum_{j \in I_{n\mathcal{P}}} x_j(t) w_j + \sum_{j \in I_{n\mathcal{Q}}} y_j(t) w_j \quad f.a.a. \ t \in \mathcal{I} \tag{4.14}$$

solves (4.7) and $u_n \in W_n$.

PROOF:
For more details on the (generalized) derivative $x'(t)$ we refer to Appendix A.2. Let (x, y) be a solution to (4.13) and define u_n as in (4.14). We can write

$$u_n(t) = \mathcal{P}u_n(t) + \mathcal{Q}u_n(t)$$

and have

$$\mathcal{D}u_n(t) = \sum_{j \in I_{n\mathcal{P}}}^{n} x_j(t) \mathcal{D}w_j.$$

Clearly $\mathcal{P}u_n \in C(\mathcal{I}, V_n)$ and so $\mathcal{P}u_n \in L_2(\mathcal{I}, V_n)$. We have $\mathcal{Q}u_n \in L_2(\mathcal{I}, V_n)$ because

$$\|\mathcal{Q}u_n\|_{L_2(\mathcal{I}, V_n)} = \left\| \sum_{j \in I_{n\mathcal{Q}}} y_j w_j \right\|_{L_2(\mathcal{I}, V_n)} \le \sum_{j \in I_{n\mathcal{Q}}} \|w_j\|_V \|y_j\|_{L_2(\mathcal{I}, \mathbb{R}^{n\mathcal{Q}})}.$$

using Minkowskis inequality. So $u_n \in L_2(\mathcal{I}, V_n)$. From (4.13c) we see that the initial value (4.7b) is fulfilled. Furthermore from (4.13a), (4.13b) we see that

$$\sum_{j \in I_{n\mathcal{P}}} x'_j(t)(\mathcal{D}w_j \,|\, \mathcal{A}w_i)_H = \langle r(t) - \mathcal{B}(t)(u_n(t)), w_i \rangle_V$$

for all $i \in \{1, \dots, n\}$. So with the identification $H_n = H_n^*$ and the properties of the derivatives $x'_j(t)$ we see that

$$[\mathcal{D}u_n(t)]' = \sum_{j \in I_{n\mathcal{P}}} x'_j(t)\mathcal{D}w_j$$

and the coordinate free Galerkin equations (4.7a) are fulfilled. Remember that we equipped H_n with the scalar product of H. We see for $j \in I_{n\mathcal{P}}$ that

$$
\begin{aligned}
([\mathcal{D}u_n(t)]' \,|\, \mathcal{A}w_j)_H &= \langle r(t) - \mathcal{B}(t)(u_n(t)), w_j \rangle_V \\
&\leq c_1 \left(\|r(t)\|_{V^*} + g(t) + \|u_n(t)\|_V \right) \|\mathcal{P}w_j\|_V
\end{aligned}
$$

for a constant $c_1 > 0$. Applying Lemma 4.5 yields

$$\|w_j\|_V \leq c_2 \|\mathcal{D}w_j\|_H, \quad j \in I_{n\mathcal{P}}$$

for a constant $c_2 > 0$ noting that the norms $\|\cdot\|_Z$ and $\|\cdot\|_H$ are equivalent on the finite dimensional space H_n. So there exists a constant $c_3 > 0$ such that

$$(\mathcal{T}_H [\mathcal{D}u_n(t)]' \,|\, \mathcal{D}w_i)_H \leq c_3 \left(\|r(t)\|_{V^*} + g(t) + \|u_n(t)\|_V \right) \|\mathcal{D}w_i\|_H.$$

for all $i \in \{1, \dots, n\}$. For $i \in I_{n\mathcal{Q}}$ is trivial. Since $[\mathcal{D}u_n(t)]' \in H_n$ we can represent it by a linear combination of the $\mathcal{D}w_i$. We obtain

$$
\begin{aligned}
c \left\| [\mathcal{D}u_n(t)]' \right\|_H^2 &\leq (\mathcal{T}_H [\mathcal{D}u_n(t)]' \,|\, [\mathcal{D}u_n(t)]')_H \\
&\leq c_4 \left(\|r(t)\|_{V^*} + g(t) + \|u_n(t)\|_V \right) \left\| [\mathcal{D}u_n(t)]' \right\|_H
\end{aligned}
$$

for a $c, c_4 > 0$ because $\mathcal{T}_H$ is strongly positive, cf. Lemma 4.8 (iv). Hence there is a $C > 0$ such that

$$\left\| [\mathcal{D}u_n(t)]' \right\|_H \leq C \left(\|r(t)\|_{V^*} + g(t) + \|u_n(t)\|_V \right)$$

The right hand side is a function in $L_2(\mathcal{I})$ and so we have that

$$[\mathcal{D}u_n]' \in L_2(\mathcal{I}, H_n).$$

and conclude $u_n \in W_n$. □

The solvability of the system (4.13) will be shown in the next lemma. Applying Lemma 4.20 the solution can be transformed into a solution to the Galerkin equations (4.7).

Lemma 4.21.
Let Assumptions 4.1, 4.3, 4.4, 4.7, 4.14 and 4.17 be fulfilled. Then the system (4.13)
has a solution (x,y) *with* $x \in C(\mathcal{I}, \mathbb{R}^{n_{\mathcal{P}}})$, $y \in L_2(\mathcal{I}, \mathbb{R}^{n_{\mathcal{Q}}})$ *and the derivative* $x'(t)$ *exists*
f.a.a. $t \in \mathcal{I}$.

PROOF:
The proof will proceed in three steps.
Step 1. Show an a priori estimate for x.
Step 2. Solve equation (4.13b) for y and insert it into equation (4.13a) to obtain an
initial value problem for x.
Step 3. Solve the initial value problem with the Theorem of Carathéodory and extend
the solution to a global solution on $\mathcal{I}$, see Appendix A.2.
Ad (1). We define the norm

$$\|x\|_{n_{\mathcal{P}},\mathcal{D}} := \left\| \sum_{j \in I_{n_{\mathcal{P}}}} x_j \mathcal{D} w_j \right\|_H$$

on $\mathbb{R}^{n_{\mathcal{P}}}$. The norm properties are inherited from the H-norm. We just have to check
that $\|x\|_{n_{\mathcal{P}},\mathcal{D}} = 0$ implies $x = 0$. If $\|x\|_{n_{\mathcal{P}},\mathcal{D}} = 0$ then $\mathcal{D}(\sum_{j \in I_{n_{\mathcal{P}}}} x_j w_j) = 0$ because $\mathcal{D}$ is
linear. Then $w := \sum_{j \in I_{n_{\mathcal{P}}}} x_j w_j \in \ker D = \operatorname{im} \mathcal{Q}$ but also $w \in \operatorname{im} \mathcal{P}$. This means $w = 0$
and thus $x = 0$ because the w_j are linearly independent. Let (x,y) be a solution to
(4.13) and u_n defined as in (4.14). $u_n \in W_n$ solves (4.7) because of Lemma 4.20. Due to
Lemma 4.18 (iii) we have that

$$\|x(t)\|_{n_{\mathcal{P}},\mathcal{D}} = \left\| \sum_{j \in I_{n_{\mathcal{P}}}} x_j(t) \mathcal{D} w_j \right\|_H = \|\mathcal{D} u_n(t)\|_H \leq \widetilde{C}$$

for $\widetilde{C} > 0$ which is independent of t. Since all norms on $\mathbb{R}^{n_{\mathcal{P}}}$ are equivalent there is a
constant $C > 0$ such that

$$\max_{t \in \mathcal{I}} \|x(t)\| \leq C. \tag{4.15}$$

Ad (2). We now solve the algebraic equation

$$b_{\mathcal{Q}}(x,y,t) = r_{\mathcal{Q}}(t)$$

having $x \in \mathbb{R}^{n_{\mathcal{P}}}$ with $\|x\| \leq C$. We define

$$B_{\mathcal{Q}}(t) : \mathbb{R}^{n_{\mathcal{Q}}} \to \mathbb{R}^{n_{\mathcal{Q}}}, \quad B_{\mathcal{Q}}(t)(y) := b_{\mathcal{Q}}(x,y,t), \quad t \in \mathcal{I}.$$

Then $B_{\mathcal{Q}}(t)$ is continuous, strongly monotone on $\mathbb{R}^{n_{\mathcal{Q}}}$ and the map $t \mapsto B_{\mathcal{Q}}(t)(y)$ is
measurable for all $y \in \mathbb{R}^{n_{\mathcal{Q}}}$. This follows directly from Lemma 4.19. Furthermore there
exists $d > 0$ and $h \in L_2(\mathcal{I})$ such that

$$\|B_{\mathcal{Q}}(t)(y)\| \leq d\|y\| + h(t) \quad \forall t \in \mathcal{I}, \ y \in \mathbb{R}^{n_{\mathcal{Q}}}.$$

This is a direct consequence from the growth condition for $\mathcal{B}(t)$:

$$
\begin{aligned}
\|b_{\mathcal{Q}}(x,y,t)\| &= \left\| \left(\langle \mathcal{B}(t)(\sum_{j \in I_{n\mathcal{P}}} x_j w_j + \sum_{j \in I_{n\mathcal{Q}}} y_j w_j), w_i \rangle_V \right)_{i \in I_{n\mathcal{Q}}} \right\| \\
&\leq \bar{c} \max_{i \in I_{n\mathcal{Q}}} (\|w_i\|_V) \left(g(t) + c \left\| \sum_{j \in I_{n\mathcal{P}}} x_j w_j + \sum_{j \in I_{n\mathcal{Q}}} y_j w_j \right\|_V \right) \\
&\leq \tilde{c} \left(g(t) + \|x\|_{n\mathcal{P}} + \|y\|_{n\mathcal{Q}} \right) \\
&\leq h(t) + d \|y\|
\end{aligned}
$$

for constants $\bar{c}, c, \tilde{c} > 0$ because all norms are equivalent on $\mathbb{R}^{n\mathcal{P}}$ and on $\mathbb{R}^{n\mathcal{Q}}$ and $\|x\| \leq C$. Fixing $t \in \mathcal{I}$ we can apply Theorem 3.6 and hence $B_{\mathcal{Q}}^{-1}(t)$ exists and is Lipschitz continuous. Consider now the operator

$$
B_{\mathcal{Q}} : L_2(\mathcal{I}, \mathbb{R}^{n\mathcal{Q}}) \to L_2(\mathcal{I}, \mathbb{R}^{n\mathcal{Q}})^*, \quad (B_{\mathcal{Q}} y)(t) := B_{\mathcal{Q}}(t)(y(t)).
$$

We see that

$$
L_2(\mathcal{I}, \mathbb{R}^{n\mathcal{Q}})^* = L_2(\mathcal{I}, (\mathbb{R}^{n\mathcal{Q}})^*) = L_2(\mathcal{I}, \mathbb{R}^{n\mathcal{Q}})
$$

and that $B_{\mathcal{Q}}$ is well-defined. This is due to the growth estimate:

$$
\|B_{\mathcal{Q}} y\|_{L_2(\mathcal{I}, \mathbb{R}^{n\mathcal{Q}})} = \int_{\mathcal{I}} \|b_{\mathcal{Q}}(x, y(t), t)\|^2 \, dt \leq \int_{\mathcal{I}} 2(h(t)^2 + d^2 \|y(t)\|^2) dt < \infty
$$

for $y \in L_2(\mathcal{I}, \mathbb{R}^{n\mathcal{Q}})$. Note that the operator family $B_{\mathcal{Q}}(t)$ fulfills all the conditions made for $\mathcal{B}(t)$ in Assumption 4.14, because the continuity of $B_{\mathcal{Q}}(t)$ implies the hemicontinuity (on $\mathbb{R}^{n\mathcal{Q}}$) and measurability is given with Lemma 4.19 (iv). As of the arguments in Lemma 4.15 (applied to $B_{\mathcal{Q}}(t)$) we see that $B_{\mathcal{Q}}$ is strongly monotone and hemicontinuous. Furthermore the inverse operator $B_{\mathcal{Q}}^{-1} : L_2(\mathcal{I}, \mathbb{R}^{n\mathcal{Q}}) \to L_2(\mathcal{I}, \mathbb{R}^{n\mathcal{Q}})$ exists and is Lipschitz continuous, cf. Theorem A.14. So the inverse operator takes the form

$$
(B_{\mathcal{Q}}^{-1} f)(t) = B_{\mathcal{Q}}^{-1}(t)(f(t)) \quad \forall t \in \mathcal{I}, \ f \in L_2(\mathcal{I}, \mathbb{R}^{n\mathcal{Q}}).
$$

Applying this gives

$$
y(t) = (B_{\mathcal{Q}}^{-1} r_{\mathcal{Q}})(t) = B_{\mathcal{Q}}^{-1}(t)(r_{\mathcal{Q}}(t)) =: b_{\mathcal{Q}}^{-1}(x, r_{\mathcal{Q}}(t), t) =: \psi_{\mathcal{Q}}(x, t)
$$

and for fixed x we see $\psi_{\mathcal{Q}}(x, \cdot) \in L_2(\mathcal{I}, \mathbb{R}^{n\mathcal{Q}})$ because $r_{\mathcal{Q}} \in L_2(\mathcal{I}, \mathbb{R}^{n\mathcal{Q}})$ and $y = B_{\mathcal{Q}}^{-1} r_{\mathcal{Q}}$. We also see that $\psi_{\mathcal{Q}}(x, t)$ is continuous w.r.t. x as follows. We fix $t \in \mathcal{I}$, let $(x_m) \subseteq \mathbb{R}^{n\mathcal{P}}$ be a sequence with $x_m \to x \in \mathbb{R}^{n\mathcal{P}}$ as $m \to \infty$ and set

$$
y := \psi_{\mathcal{Q}}(x, t), \quad y_m := \psi_{\mathcal{Q}}(x_m, t).
$$

Hence

$$b_Q(x, y, t) = r_Q(t) = b_Q(x_m, y_m, t)$$

and using the strong monotonicity of b_Q again we observe

$$\|y_m - y\| \leq \widetilde{c}\,\|b_Q(x_m, y_m, t) - b_Q(x_m, y, t)\|$$
$$= \widetilde{c}\,\|b_Q(x, y, t) - b_Q(x_m, y, t)\|$$

for a $\widetilde{c} > 0$. Finally

$$\|\psi_Q(x_m, t) - \psi_Q(x, t)\| \leq \widetilde{c}\,\|b_Q(x, y, t) - b_Q(x_m, y, t)\| \to 0 \text{ as } m \to \infty$$

because b_Q is continuous w.r.t. x.

Ad (3). We insert the expression for $y(t)$ into equation (4.13a) and obtain

$$x'(t) = G^{-1}\left(r_Q(t) - b_P(x(t), \psi_Q(x(t), t), t)\right) =: f(x(t), t)$$

because G is invertible due to Lemma 4.19. With the initial condition $x(t_0) = x_0$ this is an initial value problem on $\mathcal{I}$ which we will solve globally with Theorem A.3. In the first step of this proof we have shown the a priori estimate for x on $\mathcal{I}$ with a constant $C > 0$. It still remains to show that f fulfills a growth condition and the Carathéodory conditions on $K \times \mathcal{I}$ where $K := \left\{ x \in \mathbb{R}^{n_P} \middle| \|x\| \leq 2C \right\}$.

Growth condition. Let $\overline{y}(t) := b_Q^{-1}(\overline{x}, r_Q(t), t)$ with $\overline{x} \in K$. Strong monotonicity gives

$$\left\|b_Q^{-1}(\overline{x}, r_Q(t), t)\right\| \leq \widetilde{c}\,\|b_Q(\overline{x}, \overline{y}(t), t) - b_Q(\overline{x}, 0, t)\| \leq \overline{c}\left(\|r_Q(t)\|_\infty + \|b_Q(\overline{x}, 0, t)\|\right)$$

for a $\overline{c} > 0$. We already know that $\|b_Q(\overline{x}, 0, t)\| \leq \overline{h}(t)$ with $\overline{h} \in L_2(\mathcal{I})$ and $\overline{x} \in K$ using the growth condition of $\mathcal{B}(t)$. So for

$$k(t) := \overline{c}\left(\|r_Q(t)\| + \overline{h}(t)\right)$$

we have $k \in L_2(\mathcal{I})$ and

$$\|\psi_Q(\overline{x}, t)\| = \left\|b_Q^{-1}(\overline{x}, r_Q(t), t)\right\| \leq k(t) \text{ for all } (\overline{x}, t) \in K \times \mathcal{I} \tag{4.16}$$

Applying the growth condition to b_P we obtain

$$\|f(x, t)\| \leq \left\|G^{-1}\right\|_* \left(\|r_P(t)\| + \|b_P(x, \psi_Q(x, t), t)\|\right)$$
$$\leq \left\|G^{-1}\right\|_* \left(\|r_P(t)\| + \widetilde{c}\left(g(t) + c\left\|\sum_{j \in I_{n_P}} x_j w_j + \sum_{j \in I_{n_P}} \psi_Q(x, t) w_j\right\|_V\right)\right)$$
$$\leq \left\|G^{-1}\right\|_* \left(\|r_P(t)\| + \widetilde{c}(g(t) + c_1\|x\| + c_2\|\psi_Q(x, t)\|_\infty)\right)$$
$$\leq c_3\left(\|r_P(t)\| + g(t) + k(t)\right) =: M(t)$$

for constants $c, \tilde{c}, c_1, c_2, c_3 > 0$. Clearly $M \in L_2(\mathcal{I}) \subseteq L_1(\mathcal{I})$.

Carathéodory conditions. b_P is measurable w.r.t. t and continuous w.r.t. y. Since $\psi_Q(x, t)$ is measurable w.r.t. t it follows that from the substitution principle for measurable functions that

$$t \mapsto b_P(x, \psi_Q(x, t), t)$$

is measurable, cf. e.g. [Zei90b, p.1013]. So f is measurable w.r.t. t for all $x \in K$ because r_Q is also measurable. f is continuous w.r.t. x because b_P is continuous w.r.t. x, y (Lemma 4.19) and ψ_Q is continuous w.r.t. x as shown above.

We can now apply Theorem A.3 and so there is a solution $x_* : \mathcal{I} \to \mathbb{R}^{n_P}$ to

$$x_*'(t) = f(x_*(t), t), \quad x_*(t_0) = x_0$$

in the sense of Carathéodory, i.e. $x_* \in C(\mathcal{I}, \mathbb{R}^{n_P})$ and the derivative $x_*'(t)$ exists f.a.a. $t \in \mathcal{I}$. We set

$$y_*(t) := \psi_Q(x_*(t), t)$$

and y_* is measurable, cf. [Zei90b, p.1013]. Since $\|x_*(t)\| \leq C$ for all $t \in \mathcal{I}$ we see with (4.16) that $y_* \in L_2(\mathcal{I}, \mathbb{R}^{n_Q})$. Clearly (x_*, y_*) solves (4.13). $\qquad\square$

Unique solvability can be now concluded.

Theorem 4.22 (Unique solvability of the Galerkin equations).
Let Assumptions 4.1, 4.3, 4.4, 4.7, 4.14 and 4.17 be fulfilled. Then the Galerkin equations (4.7) have a unique solution $u_n \in W_n$.

PROOF:
Let (x, y) be a solution on $\mathcal{I}$ to (4.13) from Lemma 4.21. Then Lemma 4.20 gives a solution $u_n \in W_n$ of (4.7) and u_n is unique because of Lemma 4.18. $\qquad\square$

Remark 4.23.
For proving the unique solvability of (4.13) or (4.7) we relied on the structural assumptions made for (4.4). Note also that a global solution (on $\mathcal{I}$) was obtained. Especially considering the semi-explicit form (4.13) and knowing that b_Q is solvable w.r.t. y we have a DAE of Index 1 character. However, we cannot formally apply the definition of the Tractability Index (or ADAE Index) because it requires more smoothness w.r.t. t. Thus an application of well-known solvability results, see [GM86, LMT13, KM06], is not possible. In the linear case a solvability result for DAEs with measurable solutions is given in [LMT13, chapter 2.11.1].

4.3. Unique solvability

Theorem 4.24 (Unique solvability of (4.4)).
Let Assumptions 4.1, 4.3, 4.4, 4.7, 4.14 and 4.17 be fulfilled. Then (4.4) has a unique solution $u \in W^1_{2,\mathcal{D}}(\mathcal{I}; V, Z, H)$. Furthermore the sequence (u_n) of solutions to the Galerkin equations (4.7) converges weakly to u in $L_2(\mathcal{I}, V)$, i.e.

$$u_n \rightharpoonup u \text{ in } L_2(\mathcal{I}, V) \quad as \ n \to \infty.$$

PROOF:
We first outline the strategy of the proof.
Step 1. To increase the legibility we set

$$X := L_2(\mathcal{I}, V), \quad X^* = L_2(\mathcal{I}, V^*).$$

The spaces X, X^* and H are all reflexive, cf. [Zei90a, Proposition 23.7]. The Galerkin equations are uniquely solvable by Theorem 4.22 and a priori estimates of the solutions u_n are given in Lemma 4.18. The Theorem of Eberlein and Šmuljan states that each bounded sequence in a reflexive Banach space has a weakly convergent subsequence, cf. [Zei90a, Theorem 21.D]. Therefore a subsequence of (u_n) exists, which we again denote by (u_n), and $u \in X$, $w \in X^*$, $z_T \in H$ such that

$$u_n \rightharpoonup u \text{ in } X, \quad \mathcal{B}(u_n) \rightharpoonup w \text{ in } X^*, \quad \mathcal{D}u_n(T) \rightharpoonup z_T \text{ in } H \qquad (4.17)$$

as $n \to \infty$. According to Assumption 4.17 we have

$$z_{n0} \to z_0 \text{ in } H \text{ as } n \to \infty$$

Step 2. We show the following steps.
(2.I) The key equation

$$\phi(T)(z_T | \mathcal{A}v)_H - \phi(t_0)(z_0 | \mathcal{A}v)_H = \int_{\mathcal{I}} \langle r(t) - w(t), v \rangle_V \phi(t) + (\mathcal{A}v | \mathcal{D}u(t))_H \phi'(t) \mathrm{d}t \quad (4.18)$$

holds for all $\phi \in C^\infty(\mathcal{I})$, $v \in V$.
(2.II) The limits u, w, z_T satisfy

$$\langle [\mathcal{D}u(t)]', \mathcal{A}v \rangle_Z + \langle w(t), v \rangle_V = \langle r(t), v \rangle_V \quad \forall v \in V, \text{ f.a.a. } t \in \mathcal{I} \qquad (4.19a)$$

$$\mathcal{D}u(t_0) = z_0, \quad \mathcal{D}u(T) = z_T, \qquad (4.19b)$$

$$u \in W^1_{2,\mathcal{D}}(\mathcal{I}; V, Z, H). \qquad (4.19c)$$

(2.III) The limit elements u, w satisfy $\mathcal{B}(u) = w$. So we can conclude that $u \in W^1_{2,\mathcal{D}}$ is a solution to (4.4).

Step 3. Weak convergence of the total Galerkin sequence (u_n) in the space X to the solution u. According to Step 2 it always follows from the convergence

$$u_{n'} \rightharpoonup \overline{u} \text{ in } X \quad \text{as } n \to \infty$$

of an arbitrary subsequence of (u_n) that $\overline{u}$ is a solution of the original problem (4.4). Due to Lemma 4.16 the solution is unique and we obtain $u = \overline{u}$. Applying [Zei90a, Proposition 21.23] directly gives the convergence of the total sequence (u_n), i.e.

$$u_n \rightharpoonup u \text{ in } X \quad \text{as } n \to \infty.$$

This completes the outline of the proof and it remains to prove (2.I)-(2.III).

Ad (2.I). Let be $\phi \in C^\infty(\mathcal{I})$ and $v \in V_k$, $k \in \mathbb{N}$ fixed. Let $n \geq k$. Since $u_n \in W^1_{2,\mathcal{D}}$ it is $\mathcal{D}u_n \in W^1_2$ due to Proposition 4.2 (ii) and $\phi \mathcal{A}v \in W^1_2$ using Lemma 4.11. So an application of the integration by parts formula A.10 yields

$$(\mathcal{D}u_n(T)\,|\,\phi(T)\mathcal{A}v)_H - (\mathcal{D}u_n(t_0)\,|\,\phi(t_0)\mathcal{A}v)_H$$
$$= \int_{\mathcal{I}} \langle [\mathcal{D}u_n(t)]'\,,\phi(t)\mathcal{A}v\rangle_Z + \langle [\phi(t)\mathcal{A}v]'\,,\mathcal{D}u_n(t)\rangle_Z \mathrm{d}t$$
$$= \int_{\mathcal{I}} \langle [\mathcal{D}u_n(t)]'\,,\mathcal{A}v\rangle_Z \phi(t) + (\mathcal{A}v\,|\,\mathcal{D}u_n(t))_H \phi'(t)\mathrm{d}t$$

For obtaining the last line we used Lemma 4.11 and the continuous embedding $H \subseteq Z^*$. Since $v \in V_k \subseteq V_n$ the Galerkin equations (4.7a) can be applied and we obtain

$$\int_{\mathcal{I}} \langle [\mathcal{D}u_n(t)]'\,,\mathcal{A}v\rangle_Z \phi(t)\mathrm{d}t = \int_{\mathcal{I}} \langle r(t) - \mathcal{B}(t)(u_n(t)), v\rangle_V \phi(t)\mathrm{d}t$$
$$= \langle r - \mathcal{B}(u_n), \phi v\rangle_X$$

This gives

$$(\mathcal{D}u_n(T)\,|\,\phi(T)\mathcal{A}v)_H - (z_{n0}\,|\,\phi(t_0)\mathcal{A}v)_H = \langle r - \mathcal{B}(u_n), \phi v\rangle_X + \int_{\mathcal{I}} (\mathcal{A}v\,|\,\mathcal{D}u_n(t))_H \phi'(t)\mathrm{d}t$$

We analyze the different terms and observe that

$$\mathcal{D}u_n(T) \mapsto (\mathcal{D}u_n(T)\,|\,\phi(T)\mathcal{A}v)_H \in H^*$$
$$\mathcal{B}(u_n) \mapsto \langle \mathcal{B}(u_n), \phi v\rangle_X \in X^{**}$$

because

$$|(\mathcal{D}u_n(T)\,|\,\phi(T)\mathcal{A}v)_H| \leq \|\mathcal{D}u_n(T)\|_H \|\phi(T)\mathcal{A}v\|_H$$

and with the Hölder inequality A.7 it follows

$$|\langle \mathcal{B}(u_n), \phi v\rangle_X| \leq \int_{\mathcal{I}} |\langle \mathcal{B}(t)(u_n(t)), \phi(t)v\rangle_V|\,\mathrm{d}t \leq \|\mathcal{B}(u_n)\|_{X^*} \|\phi v\|_X.$$

Furthermore we have

$$u_n \mapsto \int_{\mathcal{I}} (\mathcal{A}v \,|\, \mathcal{D}u_n(t))_H \phi'(t) \mathrm{d}t \in X^*$$

because the linearity is apparent and

$$\left| \int_{\mathcal{I}} (\mathcal{A}v \,|\, \mathcal{D}u_n(t))_H \phi'(t)\mathrm{d}t \right| \leq c_1 \left\| \mathcal{A}v \right\|_Z \left\| \phi' \right\|_\infty \int_{\mathcal{I}} \left\| \mathcal{D}u_n(t) \right\|_Z \mathrm{d}t$$

$$\leq c_2 \int_{\mathcal{I}} \left\| u_n(t) \right\|_V \mathrm{d}t \leq c_3 \left\| u_n \right\|_{L_2(\mathcal{I},V)}$$

using that $Z \subseteq H$ is continuous and the Hölder inequality A.7. Letting $n \to \infty$ but keeping $v \in V_k$ fixed we obtain

$$(\mathcal{D}u_n(T) \,|\, \phi(T)\mathcal{A}v)_H \to (z_T \,|\, \phi(T)\mathcal{A}v)_H,$$
$$(z_{n0} \,|\, \phi(t_0)\mathcal{A}v)_H \to (z_0 \,|\, \phi(t_0)\mathcal{A}v)_H,$$
$$\langle \mathcal{B}(u_n), \phi v \rangle_X \to \langle w, \phi v \rangle_X,$$
$$\int_{\mathcal{I}} (\mathcal{A}v \,|\, \mathcal{D}u_n(t))_H \phi'(t)\mathrm{d}t \to \int_{\mathcal{I}} (\mathcal{A}v \,|\, \mathcal{D}u(t))_H \phi'(t)\mathrm{d}t.$$

recalling that $z_{n0} \to z_0$ and the weak convergence of Step 1. We conclude

$$(z_T \,|\, \phi(T)\mathcal{A}v)_H - (z_0 \,|\, \phi(t_0)\mathcal{A}v)_H = \langle r - w, \phi v \rangle_X + \int_{\mathcal{I}} (\mathcal{A}v \,|\, \mathcal{D}u(t))_H \phi'(t)\mathrm{d}t$$

for all $v \in \bigcup_k V_k$. The set $\bigcup_k V_k$ is dense in V because the V_k are a Galerkin scheme. Let now $v \in V$ be arbitrary then there is a sequence $(v_m) \subseteq \bigcup_k V_k$ such that $v_m \to v$ in V as $m \to \infty$. We have

$$\left\| \mathcal{A}v_m - \mathcal{A}v \right\|_H \leq c \left\| \mathcal{T} \right\|_{L(Z)} \left\| \mathcal{D} \right\|_{L(V,Z)} \left\| v - v_m \right\|_V$$

for a $c > 0$ and so $\mathcal{A}v_m \to \mathcal{A}v$ as $m \to \infty$. Additionally

$$\langle r - w, \phi(v_m - v) \rangle_X \leq \left\| r - w \right\|_{X^*} \left\| \phi(v_m - v) \right\|_X$$
$$\leq \widetilde{c} \left\| r - w \right\|_{X^*} \left\| \phi \right\|_\infty \left\| v_m - v \right\|_V \to 0 \text{ as } m \to \infty$$

with $\widetilde{c} > 0$. Hence we obtain the key equation (4.18) for all $v \in V$.
Ad (2.II). Applying the key equation (4.18) for $\phi \in C_0^\infty(\mathcal{I})$, $v \in V$:

$$- \int_{\mathcal{I}} (\mathcal{D}u(t) \,|\, \mathcal{A}v)_H \phi'(t)\mathrm{d}t = \int_{\mathcal{I}} \langle r(t) - w(t), v \rangle_V \phi(t)\mathrm{d}t. \tag{4.20}$$

For $v \in \ker \mathcal{D} = \operatorname{im} \mathcal{Q}$ we obtain

$$\int_{\mathcal{I}} \langle r(t) - w(t), v \rangle_V \phi(t)\mathrm{d}t = 0 \quad \text{f.a.a. } t \in \mathcal{I}.$$

The mapping

$$t \mapsto \langle r(t) - w(t), v \rangle_V \phi(t)$$

is integrable and hence an application of the well-known variational lemma reveals

$$\langle r(t) - w(t), v \rangle_V = 0 \quad \forall v \in \operatorname{im} \mathcal{Q} \quad \text{f.a.a. } t \in \mathcal{I}. \tag{4.21}$$

$\mathcal{A}$ is surjective because $\mathcal{T}$ and $\mathcal{D}$ are. So for $z \in Z$ there is a unique $v \in \operatorname{im} \mathcal{P}$ such that $z = \mathcal{A}v$. The following definition is well-defined for all $z \in Z$:

$$\langle \bar{z}(t), z \rangle_Z := \langle r(t) - w(t), v \rangle_V \quad \text{f.a.a. } t \in \mathcal{I} \tag{4.22}$$

with $z = \mathcal{A}v$, $v \in \operatorname{im} \mathcal{P} = \ker \mathcal{Q}$. We show $\bar{z} \in L_2(\mathcal{I}, Z^*)$. There is a $c > 0$ such that

$$\|v\|_V \le c \|\mathcal{A}v\|_Z \quad \forall v \in \ker \mathcal{Q}$$

because $\mathcal{T}$ is bijective and Lemma 4.5. Since $\mathcal{A}v = 0$ for $v \in \operatorname{im} \mathcal{Q}$ it is

$$\langle \bar{z}(t), \mathcal{A}v \rangle_Z = 0 \quad \forall v \in \operatorname{im} \mathcal{Q} \tag{4.23}$$

and it follows that

$$
\begin{aligned}
\|\bar{z}(t)\|_{Z^*} &= \sup_{\|z\|_Z \le 1} |\langle \bar{z}(t), z \rangle_Z| \\
&= \sup_{\|\mathcal{A}v\|_Z \le 1} |\langle \bar{z}(t), \mathcal{A}v \rangle_Z| \\
&= \sup_{\substack{\|\mathcal{A}v\|_Z \le 1 \\ v \in \ker \mathcal{Q}}} |\langle \bar{z}(t), \mathcal{A}v \rangle_Z| \\
&\le \sup_{\substack{\|\mathcal{A}v\|_Z \le 1 \\ v \in \ker \mathcal{Q}}} |\langle r(t) - w(t), v \rangle_V| \\
&\le \sup_{\substack{\|\mathcal{A}v\|_Z \le 1 \\ v \in \ker \mathcal{Q}}} \|r(t) - w(t)\|_{V^*} \|v\|_V \\
&\le \tilde{c} \sup_{\substack{\|\mathcal{A}v\|_Z \le 1 \\ v \in \ker \mathcal{Q}}} \|r(t) - w(t)\|_{V^*} \|\mathcal{A}v\|_Z
\end{aligned}
$$

f.a.a $t \in \mathcal{I}$ which implies

$$\|\bar{z}(t)\|_{Z^*} \le C \|r(t) - w(t)\|_{V^*}$$

for constants $\tilde{c}$, $C > 0$. Since r, $w \in L_2(\mathcal{I}, V^*)$ we have $\bar{z} \in L_2(\mathcal{I}, Z^*)$. With (4.20) and Proposition A.9 we can conclude that

$$\bar{z} = [\mathcal{D}u]' \in L_2(\mathcal{I}, Z^*).$$

Since $u \in X$ we finally have $u \in W^1_{2,\mathcal{D}}$. With (4.21), (4.22) and (4.23) it holds for all $v \in V$ that

$$\langle r(t) - w(t), v \rangle_V = \langle \overline{z}(t), \mathcal{A}v \rangle_Z = \langle [\mathcal{D}u]'(t), \mathcal{A}v \rangle_Z \quad \text{f.a.a. } t \in \mathcal{I}$$

and thus (4.19a), (4.19c) are fulfilled.

It remains to show (4.19b). Since $u \in W^1_{2,\mathcal{D}}$ we have $\mathcal{D}u \in W^1_2$ and $\phi \mathcal{A}v \in W^1_2$ for all $v \in V$ and $\phi \in C_0^\infty(\mathcal{I})$. We can apply the integration by parts formula (Proposition 4.2) and obtain

$$(\mathcal{D}u(T) \,|\, \phi(T)\mathcal{A}v)_H - (\mathcal{D}u(t_0) \,|\, \phi(t_0)\mathcal{A}v)_H$$

$$= \int_{\mathcal{I}} \langle [\mathcal{D}u]'(t), \phi(t)\mathcal{A}v \rangle_Z + \langle [\phi\mathcal{A}v]'(t), \mathcal{D}u(t) \rangle_Z \mathrm{d}t$$

$$\overset{(4.19a)}{=} \int_{\mathcal{I}} \langle r(t) - w(t), v \rangle_V \phi(t) + (\mathcal{A}v \,|\, \mathcal{D}u(t))_H \phi'(t) \mathrm{d}t$$

$$\overset{(4.18)}{=} \phi(T)(z_T \,|\, \mathcal{A}v)_H - \phi(t_0)(z_0 \,|\, \mathcal{A}v)_H.$$

Choosing $\phi \in C^\infty(\mathcal{I})$ with $\phi(t_0) = 0$ and $\phi(T) = 1$ gives

$$(\mathcal{D}u(t_0) - z_0 \,|\, \mathcal{A}v)_H = 0 \quad \forall v \in V.$$

Since $\mathcal{A}$ is surjective we have

$$(\mathcal{D}u(t_0) - z_0 \,|\, z)_H = 0 \quad \forall z \in Z$$

and it follows $\mathcal{D}u(t_0) = z_0$ because $Z \subseteq H$ dense. With the same arguments it follows $\mathcal{D}u(T) = z_T$ taking a $\phi \in C^\infty(\mathcal{I})$ with $\phi(t_0) = 1$ and $\phi(T) = 0$.

Ad (2.III). It is $u_n \rightharpoonup u$ in X and $\mathcal{B}(u_n) \rightharpoonup w$ in X^* as $n \to \infty$ and $\mathcal{B}$ is monotone and hemicontinuous, cf. Lemma 4.15. By showing

$$\overline{\lim_{n \to \infty}} \langle \mathcal{B}(u_n), u_n \rangle_X \leq \langle w, u \rangle_X \tag{4.24}$$

the fundamental monotonicity trick can be applied, cf. [Zei90b, p.474 or p.778], which implies $\mathcal{B}(u) = w$. Since $u_n \in W^1_{2,\mathcal{D}}$ we have $\mathcal{D}u_n \in W^1_2$ and $\mathcal{A}u_n \in W^1_2$, cf. Lemma 4.10. Therefore we can apply the integration by parts formula 4.13 and observe

$$\frac{1}{2}\|\mathcal{S}\mathcal{D}u_n(T)\|_H - \frac{1}{2}\|\mathcal{S}\mathcal{D}u_n(t_0)\|_H = \int_{\mathcal{I}} \langle [\mathcal{D}u_n]'(t), \mathcal{A}u_n(t) \rangle_Z \mathrm{d}t$$

$$\overset{(4.7a)}{=} \int_{\mathcal{I}} \langle r(t) - \mathcal{B}(t)(u_n(t)), u_n(t) \rangle_V \mathrm{d}t.$$

Hence

$$\langle \mathcal{B}(u_n), u_n \rangle_X = \langle r, u_n \rangle_X + \frac{1}{2}\left(\|\mathcal{S}\mathcal{D}u_n(t_0)\|_H - \|\mathcal{S}\mathcal{D}u_n(T)\|_H \right).$$

Because of (4.17) we have that

$$\mathcal{SD}u_n(t_0) \to \mathcal{SD}u(t_0) \text{ in } H, \quad \mathcal{SD}u_n(T) \rightharpoonup \mathcal{SD}u(T) \text{ in } H$$

and hence

$$\|\mathcal{SD}u(T)\|_H \leq \varliminf_{n\to\infty} \|\mathcal{SD}u_n(T)\|_H \,.$$

Furthermore we have

$$\langle r, u_n \rangle_X \to \langle r, u \rangle_X \quad \text{as } n \to \infty$$

because $u_n \rightharpoonup u$ in X. We conclude

$$\varlimsup_{n\to\infty} \langle \mathcal{B}(u_n), u_n \rangle_X \leq \langle r, u \rangle_X + \frac{1}{2} \left(\|\mathcal{SD}u(t_0)\|_H - \|\mathcal{SD}u_n(T)\|_H \right).$$

Integrating by parts reveals

$$\frac{1}{2} \left(\|\mathcal{SD}u(t_0)\|_H - \|\mathcal{SD}u(T)\|_H \right) \quad = \quad - \int_{\mathcal{I}} \langle [\mathcal{D}u]'(t), \mathcal{A}u(t) \rangle_Z \mathrm{d}t$$

$$\overset{(4.19a)}{=} \quad \langle w - r, u \rangle_X .$$

Finally we get (4.24). $\qquad\qquad\qquad\qquad\qquad\qquad\qquad\qquad\qquad\qquad\qquad\qquad\qquad \square$

The strategy of the proof of Theorem 4.24 is strongly related to the one for proving unique solvability of evolution equations, cf. [Zei90b, Theorem 30.A]. As already pointed out in [Tis04] the main differences are:

- The solution u is in the space $W_{2,\mathcal{D}}^1$ and not in W_2^1.

- The Galerkin equations (4.7) represent a nonlinear DAE instead of an ODE.

- Initial conditions are given for $\mathcal{D}u(t_0)$ and not for $u(t_0)$ completely.

- Assumption 4.4 is needed to define the generalized derivative $[\mathcal{D}u]'$.

In Theorem 4.24 we have proven the unique solvability of (4.4) via the weak convergence of the solutions of the corresponding Galerkin equations (4.7). We will prove in the next theorem that this convergence is indeed strong in $L_2(\mathcal{I}, V)$. Moreover, the dynamical part of the solution $\mathcal{D}u_n$ converges in the space $C(\mathcal{I}, H)$.

Theorem 4.25 (Strong convergence of the Galerkin solutions).
Let Assumptions 4.1, 4.3, 4.4, 4.7, 4.14 and 4.17 be fulfilled. Let u be the unique solution to (4.4) and u_n the unique solution of the corresponding Galerkin equations (4.7). Then the following holds:

(i) $\mathcal{D}u_n \to \mathcal{D}u$ in $C(\mathcal{I}, H)$ as $n \to \infty$,

(ii) $u_n \to u$ in $L_2(\mathcal{I}, V)$ as $n \to \infty$.

PROOF:

(i) Let $u \in W_{2,\mathcal{D}}^1$ be the solution to (4.4) and $u_n \in W_n \subseteq W_{2,\mathcal{D}}^1$ be the solution to (4.7). The set of polynomials $p : \mathcal{I} \to V$ is dense in $W_{2,\mathcal{D}}^1$, cf. Lemma 4.6. So we find for $u \in W_{2,\mathcal{D}}^1$ and $\varepsilon > 0$ a polynomial

$$p(t) = \sum_i a_i t^i, \quad a_i \in V$$

and

$$\|u - p\|_{W_{2,\mathcal{D}}^1} = \|u - p\|_{L_2(\mathcal{I}, V)} + \left\| [\mathcal{D}u]' - [\mathcal{D}p]' \right\|_{L_2(\mathcal{I}, Z^*)} \leq \varepsilon.$$

The set $\bigcup_{n \in \mathbb{N}} V_n$ is dense in V. So the set of polynomials with coefficients in $\bigcup_{n \in \mathbb{N}} V_n$ is dense in $W_{2,\mathcal{D}}^1$. That means that there is a sequence (p_n) of polynomials $p_n : \mathcal{I} \to V_n$ with

$$p_n \to u \quad \text{in } W_{2,\mathcal{D}}^1 \quad \text{as } n \to \infty. \tag{4.25}$$

The continuity of the embedding $W_2^1 \subseteq C(\mathcal{I}, H)$ implies

$$\max_{t \in \mathcal{I}} \|\mathcal{D}u(t) - \mathcal{D}p_n(t)\|_H \leq c_1 \|\mathcal{D}u - \mathcal{D}p_n\|_{W_2^1} \leq \tilde{c}_1 \|u - p_n\|_{W_{2,\mathcal{D}}^1}$$

for constants $c_1, \tilde{c}_1 > 0$ and thus

$$\max_{t \in \mathcal{I}} \|\mathcal{D}u(t) - \mathcal{D}p_n(t)\|_H \to 0 \quad \text{as } n \to \infty. \tag{4.26}$$

Since

$$\max_{t \in \mathcal{I}} \|\mathcal{D}u_n(t) - \mathcal{D}u(t)\|_H \leq \max_{t \in \mathcal{I}} \|\mathcal{D}u_n(t) - \mathcal{D}p_n(t)\|_H + \max_{t \in \mathcal{I}} \|\mathcal{D}p_n(t) - \mathcal{D}u(t)\|_H$$

it is enough to show that

$$\max_{t \in \mathcal{I}} \|\mathcal{D}u_n(t) - \mathcal{D}p_n(t)\|_H \to 0 \quad \text{as } n \to \infty. \tag{4.27}$$

We also remark that

$$\begin{aligned} \|\mathcal{S}\mathcal{D}u_n(t_0) - \mathcal{S}\mathcal{D}p_n(t_0)\|_H &\leq \|\mathcal{S}\mathcal{D}u_n(t_0) - \mathcal{S}\mathcal{D}u(t_0)\|_H + \|\mathcal{S}\mathcal{D}u(t_0) - \mathcal{S}\mathcal{D}p_n(t_0)\|_H \\ &\leq \|\mathcal{S}z_{n0} - \mathcal{S}z_0\|_H + \max_{t \in \mathcal{I}} \|\mathcal{S}\mathcal{D}u(t) - \mathcal{S}\mathcal{D}p_n(t)\|_H \end{aligned}$$

and so

$$\|\mathcal{S}\mathcal{D}u_n(t_0) - \mathcal{S}\mathcal{D}p_n(t_0)\|_H \to 0 \quad \text{as } n \to \infty \tag{4.28}$$

because $\mathcal{S} \in L(H)$, $z_{n0} \to z_0$ in H and $\overline{(4.26)}$. We will show in the following that

$$\max_{t \in \mathcal{I}} \|\mathcal{S}\mathcal{D}u_n(t) - \mathcal{S}\mathcal{D}p_n(t)\|_H^2 - \|\mathcal{S}\mathcal{D}u_n(t_0) - \mathcal{S}\mathcal{D}p_n(t_0)\|_H^2 \to 0 \quad \text{as } n \to \infty. \quad (4.29)$$

Using the Galerkin equations (4.7a) and the original equations (4.4a) we have that

$$\langle [\mathcal{D}u_n(t)]', \mathcal{A}(u_n(t) - p_n(t)) \rangle_Z$$
$$\overset{(4.7a)}{=} \langle r(t) - \mathcal{B}(t)(u_n(t)), u_n(t) - p_n(t) \rangle_V$$
$$\overset{(4.4a)}{=} \langle [\mathcal{D}u(t)]', \mathcal{A}(u_n(t) - p_n(t)) \rangle_Z + \langle \mathcal{B}(t)(u(t)) - \mathcal{B}(t)(u_n(t)), u_n(t) - p_n(t) \rangle_V$$
$$\leq \langle [\mathcal{D}u(t)]', \mathcal{A}(u_n(t) - p_n(t)) \rangle_Z + \langle \mathcal{B}(t)(u(t)) - \mathcal{B}(t)(u_n(t)), u(t) - p_n(t) \rangle_V$$

for all $t \in \mathcal{I}$. The last line was obtained using the monotonicity of $\mathcal{B}(t)$, i.e.

$$\langle \mathcal{B}(t)(u(t)) - \mathcal{B}(t)(u_n(t)), u(t) - u_n(t) \rangle_V \geq 0.$$

Then integration by parts, cf. Corollary 4.13, gives for any $t \in \mathcal{I}$:

$$\frac{1}{2} \left(\|\mathcal{S}\mathcal{D}u_n(t) - \mathcal{S}\mathcal{D}p_n(t)\|_H^2 - \|\mathcal{S}\mathcal{D}u_n(t_0) - \mathcal{S}\mathcal{D}p_n(t_0)\|_H^2 \right)$$
$$= \int_{t_0}^t \langle [\mathcal{D}u_n(s) - \mathcal{D}p_n(s)]', \mathcal{A}(u_n(s) - p_n(s)) \rangle_Z \mathrm{d}s$$
$$\leq \int_{t_0}^t \langle [\mathcal{D}u(s)]' - [\mathcal{D}p_n(s)]', \mathcal{A}(u_n(s) - p_n(s)) \rangle_Z \mathrm{d}s$$
$$\quad + \int_{t_0}^t \langle \mathcal{B}(t)(u(t)) - \mathcal{B}(t)(u_n(t)), u(t) - p_n(t) \rangle_V \mathrm{d}s$$
$$\leq \left\| [\mathcal{D}u]' - [\mathcal{D}p_n]' \right\|_{L_2(\mathcal{I},Z^*)} \|\mathcal{A}u_n - \mathcal{A}p_n\|_{L_2(\mathcal{I},Z)}$$
$$\quad + \|\mathcal{B}(u) - \mathcal{B}(u_n)\|_{L_2(\mathcal{I},V^*)} \|u - p_n\|_{L_2(\mathcal{I},V)}.$$

The last line is a consequence of the Hölder inequality, cf. Proposition A.7. We have

$$\|\mathcal{A}u_n - \mathcal{A}p_n\|_{L_2(\mathcal{I},Z)}^2 \leq \int_{\mathcal{I}} \|\mathcal{A}\|_{L(V,Z)}^2 \|u_n(t) - p_n(t)\|_V^2 \, \mathrm{d}t \leq c_2 \|u_n - p_n\|_{L_2(\mathcal{I},V)}^2 \leq \widetilde{C}$$

for constants c_2, $\widetilde{C} > 0$ because $\|u_n - p_n\|_{L_2(\mathcal{I},V)}$ is bounded. This can be easily verified by observing that

$$\|u_n - p_n\|_{L_2(\mathcal{I},V)} \leq \|u_n\|_{L_2(\mathcal{I},V)} + \|p_n\|_{L_2(\mathcal{I},V)}$$

and $\|u_n\|_{L_2(\mathcal{I},V)} \leq C$ by the a priori estimate, cf. Lemma 4.18. Since $p_n \to u$ in $W_{2,\mathcal{D}}^1$ we have $p_n \to u$ in $L_2(\mathcal{I}, V)$, i.e. $\|p_n\|_{L_2(\mathcal{I},V)}$ is bounded. Furthermore

$$\|\mathcal{B}(u) - \mathcal{B}(u_n)\|_{L_2(\mathcal{I},V^*)} \leq \|\mathcal{B}(u)\|_{L_2(\mathcal{I},V^*)} + \|\mathcal{B}(u_n)\|_{L_2(\mathcal{I},V^*)}$$
$$\leq c(\|g\|_{L_2(\mathcal{I})} + \|u\|_{L_2(\mathcal{I},V)} + \|u_n\|_{L_2(\mathcal{I},V)})$$

is bounded because $\mathcal{B}$ satisfies the growth condition, cf. Lemma 4.15, and the a priori estimates from Lemma 4.18 hold. So we deduce that there is a $c_3 > 0$ such that

$$\|\mathcal{SD}u_n(t) - \mathcal{SD}p_n(t)\|_H^2 - \|\mathcal{SD}u_n(t_0) - \mathcal{SD}p_n(t_0)\|_H^2$$
$$\leq \ c_3 \left(\left\| [\mathcal{D}u]' - [\mathcal{D}p_n]' \right\|_{L_2(\mathcal{I},Z^*)} + \|u - p_n\|_{L_2(\mathcal{I},V)} \right)$$
$$= \ c_3 \|u - p_n\|_{W_{2,\mathcal{D}}^1} .$$

The limit (4.29) follows immediately because of (4.25). Then (4.27) follows with (4.28) and Lemma 4.8 (iii). Finally we get $\mathcal{D}u_n \to \mathcal{D}u$ in $C(\mathcal{I},H)$ as $n \to \infty$.

(ii) Set $X := L_2(\mathcal{I},V)$ and $Y := L_2(\mathcal{I},Z)$. From Theorem 4.24 we know that

$$u_n \rightharpoonup u \text{ in } X, \quad \mathcal{B}(u_n) \rightharpoonup \mathcal{B}(u) \text{ in } X^*, \quad \mathcal{D}u_n(T) \rightharpoonup \mathcal{D}u(T) \text{ in } H \quad \text{as } n \to \infty. \quad (4.30)$$

We first remark that integration by parts gives

$$\frac{1}{2} \left(\|\mathcal{SD}(u(T) - u_n(T))\|_H^2 - \|\mathcal{SD}(u(t_0) - u_n(t_0))\|_H^2 \right)$$
$$= \ \int_{\mathcal{I}} \langle [\mathcal{D}u(t)]' - [\mathcal{D}u_n(t)]', \mathcal{A}(u(t) - u_n(t)) \rangle_Z \mathrm{d}t$$
$$\overset{(4.4a)}{=} \ \langle r - \mathcal{B}(u), u - u_n \rangle_X - \langle [\mathcal{D}u_n]', \mathcal{A}(u - u_n) \rangle_Y$$

and with Lemma 4.10:

$$(\mathcal{D}u_n(T) | \mathcal{A}u(T))_H - (\mathcal{D}u_n(t_0) | \mathcal{A}u(t_0))_H$$
$$= \ \int_{\mathcal{I}} \langle [\mathcal{D}u_n(t)]', \mathcal{A}u(t) \rangle_Z + \langle [\mathcal{T}\mathcal{D}u(t)]', \mathcal{D}u_n(t) \rangle_Z \mathrm{d}t$$
$$= \ \int_{\mathcal{I}} \langle [\mathcal{D}u_n(t)]', \mathcal{A}u(t) \rangle_Z + \langle [\mathcal{D}u(t)]', \mathcal{A}u_n(t) \rangle_Z \mathrm{d}t$$
$$\overset{(4.4a)}{=} \ \langle [\mathcal{D}u_n]', \mathcal{A}u \rangle_Y + \langle r - \mathcal{B}(u), u_n \rangle_X .$$

Applying the strong monotonicity of $\mathcal{B}$, cf. Lemma 4.15, gives

$$\mu \|u - u_n\|_X^2$$
$$\leq \ \langle \mathcal{B}(u) - \mathcal{B}(u_n), u - u_n \rangle_X + \frac{1}{2} \|\mathcal{SD}(u(T) - u_n(T))\|_H^2$$
$$\leq \ \langle r - \mathcal{B}(u_n), u - u_n \rangle_X - \langle [\mathcal{D}u_n]', \mathcal{A}(u - u_n) \rangle_Y + \frac{1}{2} \|\mathcal{SD}(u(t_0) - u_n(t_0))\|_H^2$$
$$\overset{(4.7a)}{\leq} \ \langle r - \mathcal{B}(u_n), u \rangle_X - \langle [\mathcal{D}u_n]', \mathcal{A}u \rangle_Y + \frac{1}{2} \|\mathcal{SD}(u(t_0) - u_n(t_0))\|_H^2$$
$$\leq \ \langle r - \mathcal{B}(u_n), u \rangle_X + \frac{1}{2} \|\mathcal{SD}(u(t_0) - u_n(t_0))\|_H^2$$
$$\quad + \langle r - \mathcal{B}(u), u_n \rangle_X - (\mathcal{D}u_n(T) | \mathcal{A}u(T))_H + (\mathcal{D}u_n(t_0) | \mathcal{A}u(t_0))_H$$
$$\to \ 2\langle r - \mathcal{B}(u), u \rangle_X - (\mathcal{D}u(T) | \mathcal{A}u(T))_H + (\mathcal{D}u(t_0) | \mathcal{A}u(t_0))_H \quad \text{as } n \to \infty.$$

The limit follows with the same arguments as in the proof of Theorem 4.24 because of (4.30). Integration by parts reveals as above

$$2\langle r - \mathcal{B}(u), u \rangle_X \overset{(4.4a)}{=} 2\langle [\mathcal{D}u]', \mathcal{A}u \rangle_Y$$
$$= (\mathcal{D}u(T) | \mathcal{A}u(T))_H - (\mathcal{D}u(t_0) | \mathcal{A}u(t_0))_H.$$

We conclude $\|u - u_n\|_X \to 0$ as $n \to \infty$. $\qquad\qquad\qquad\square$

4.4. Dependence on the data

In this section we investigate solutions to (4.4) w.r.t. perturbations on the right hand side and perturbed initial values. More precisely, we let $\delta \in L_2(\mathcal{I}, V^*)$ be a perturbation and $z_0^\delta \in H$ be the perturbed initial value and we obtain the perturbed problem

$$\langle \mathcal{A}^* [\mathcal{D}u(t)]', v \rangle_V + \langle \mathcal{B}(t)(u(t)), v \rangle_V = \langle r(t) + \delta(t), v \rangle_V, \qquad (4.31a)$$
$$\mathcal{D}u(t_0) = z_0^\delta \in H. \qquad (4.31b)$$

for all $v \in V$ and f.a.a. $t \in \mathcal{I}$. We derive the following perturbation result.

Theorem 4.26 (Perturbation result).
Let Assumptions 4.1, 4.3, 4.4, 4.7, 4.14 and 4.17 be fulfilled and let $u \in W_{2,\mathcal{D}}^1$ be the unique solution to (4.4). Let there be a sequence $(z_{0n}^\delta) \subseteq Z$ with $z_{0n}^\delta \to z_0^\delta$ in H in analogy to Assumption 4.17 and $\delta \in L_2(\mathcal{I}, V^*)$. Then (4.31) has a unique solution $u_\delta \in W_{2,\mathcal{D}}^1$ and there exists a $C > 0$ such that

$$\|u - u_\delta\|_{L_2(\mathcal{I},V)} \leq C \left(\left\|z_0 - z_0^\delta\right\|_H + \|\delta\|_{L_2(\mathcal{I},V^*)} \right).$$

If $\mathcal{B}(t)$ is also Lipschitz continuous, i.e. there exists an $L > 0$ such that

$$\|\mathcal{B}(t)(u) - \mathcal{B}(t)(v)\|_{V^*} \leq L \|u - v\|_V \quad \forall u, v \in V, \ t \in \mathcal{I}$$

then there exists a $\widetilde{C} > 0$ such that

$$\|u - u_\delta\|_{W_{2,\mathcal{D}}^1} \leq \widetilde{C} \left(\left\|z_0 - z_0^\delta\right\|_H + \|\delta\|_{L_2(\mathcal{I},V^*)} \right).$$

PROOF:
Set $X := L_2(\mathcal{I}, V)$ and $Y := L_2(\mathcal{I}, Z)$. The unique solvability of (4.31) follows directly from Theorem 4.24. Let $u_\delta \in W_{2,\mathcal{D}}^1$ be the unique solution of (4.31) and u the one of (4.4). Taking the difference of (4.31a) and (4.4a) gives

$$\langle [\mathcal{D}u_\delta(t)]' - [\mathcal{D}u(t)]', \mathcal{A}v \rangle_Z + \langle \mathcal{B}(t)(u_\delta(t)) - \mathcal{B}(t)(u(t)), v \rangle_V = \langle \delta(t), v \rangle_V.$$

for all $v \in V$ and f.a.a. $t \in \mathcal{I}$. Using the strong monotonicity of $\mathcal{B}$, cf. Lemma 4.15, and the integration by parts formula, cf. Corollary 4.13, gives

$$
\begin{aligned}
\mu \left\| u_\delta - u \right\|_X^2 &\leq \langle \mathcal{B}(u_\delta) - \mathcal{B}(u), u_\delta - u \rangle_X \\
&= \langle \delta, u_\delta - u \rangle_X - \langle [\mathcal{D} u_\delta - \mathcal{D} u]', \mathcal{A} u_\delta - \mathcal{A} u \rangle_Y \\
&= \langle \delta, u_\delta - u \rangle_X + \frac{1}{2} \left\| \mathcal{S} \mathcal{D}(u_\delta(t_0) - u(t_0)) \right\|_H - \frac{1}{2} \left\| \mathcal{S} \mathcal{D}(u_\delta(T) - u(T)) \right\|_H \\
&\leq \left\| \delta \right\|_{X^*} \left\| u_\delta - u \right\|_X + \frac{1}{2} \left\| \mathcal{S} \mathcal{D}(u_\delta(t_0) - u(t_0)) \right\|_H .
\end{aligned}
$$

We have

$$
\left\| \delta \right\|_{X^*} \left\| u_\delta - u \right\|_X \leq \frac{2}{\mu} \left\| \delta \right\|_{X^*}^2 + \frac{\mu}{2} \left\| u_\delta - u \right\|_X^2
$$

with the classical inequality (4.8) and

$$
\left\| \mathcal{S} \mathcal{D}(u_\delta(t_0) - u(t_0)) \right\|_H \leq c_1 \left\| z_0^\delta - z_0 \right\|_H
$$

for a $c_1 > 0$ because $\mathcal{S} \in L(H)$. So there is a constant $C > 0$ such that

$$
\left\| u - u_\delta \right\|_X \leq C \left(\left\| z_0 - z_0^\delta \right\|_H + \left\| \delta \right\|_{X^*} \right). \tag{4.32}
$$

Let now $\mathcal{B}(t)$ be also Lipschitz continuous. We have from above

$$
\begin{aligned}
\langle [\mathcal{D} u_\delta(t)]' - [\mathcal{D} u(t)]', \mathcal{A} v \rangle_Z &= \langle \delta(t) - \mathcal{B}(t)(u_\delta(t)) + \mathcal{B}(t)(u(t)), v \rangle_V \\
&\leq \left(\left\| \delta(t) \right\|_{V^*} + L \left\| u_\delta(t) - u(t) \right\|_V \right) \left\| v \right\|_V
\end{aligned}
$$

for all $v \in V$. The restricted operator $\mathcal{A}|_{\operatorname{im} \mathcal{P}}$ is bijective and for $v \in \operatorname{im} \mathcal{P}$ we have $\left\| v \right\|_V \leq c_2 \left\| \mathcal{A} v \right\|_Z$ for a $c_2 > 0$, cf. Lemma 4.5. So

$$
\left\| [\mathcal{D} u_\delta(t)]' - [\mathcal{D} u(t)]' \right\|_{Z^*} \leq c_2 \left(\left\| \delta(t) \right\|_{V^*} + L \left\| u_\delta(t) - u(t) \right\|_V \right)
$$

and hence using the classical inequality (4.8) again gives

$$
\begin{aligned}
\left\| [\mathcal{D}(u_\delta - u)]' \right\|_{Y^*}^2 &= \int_{\mathcal{I}} \left\| [\mathcal{D}(u_\delta(t) - u(t))]' \right\|_{Z^*}^2 \, \mathrm{d}t \\
&\leq c_3 \left(\left\| \delta \right\|_{X^*}^2 + \left\| u_\delta - u \right\|_X^2 \right).
\end{aligned}
$$

for a $c_3 > 0$. Combining this with (4.32) gives the desired $W_{2,\mathcal{D}}^1$-estimate. $\qquad \square$

As a direct consequence of Theorem 4.26 we see that the map

$$
\varphi_{L_2} : H \times L_2(\mathcal{I}, V^*) \to L_2(\mathcal{I}, V), \quad (z_0, r) \mapsto u
$$

is continuous. If $\mathcal{B}(t)$ is also Lipschitz continuous then the map

$$\varphi_{W_{2,\mathcal{D}}^1} : H \times L_2(\mathcal{I}, V^*) \to W_{2,\mathcal{D}}^1(\mathcal{I}; V, Z, H), \quad (z_0, r) \mapsto u$$

is continuous and we have continuous dependence on the data as for the linear case, cf. [Tis04, Theorem 4.26]. Furthermore the estimates of Theorem 4.26 show that (4.4) has Perturbation Index 1, cf. Definition 2.4. This also holds for the Galerkin equations (4.7).

In chapter 2 the concept of the ADAE Index has been presented. A direct application of Definition 2.3 is not possible under the assumptions of Theorem 4.24. The unique solution $u \in W_{2,\mathcal{D}}^1$ is not necessarily continuously differentiable. Moreover, the Fréchet-differentiability of the operator $\mathcal{B}(t)$ is not assumed to prove existence. However, if we assume Fréchet-differentiability of $\mathcal{B}(t)$, it is possible to define operators $\mathcal{G}_0$ and $\mathcal{G}_1$ similar to the ones in Definition 2.3. With these operators we are able to characterize the index behavior by proving the required properties in definition 2.3.

First we mention that the strong monotonicity of $\mathcal{B}(t)$ carries over to the Fréchet-derivative $\mathcal{B}_0$.

Lemma 4.27.
Let $\mathcal{B} : V \to V^*$ be Fréchet differentiable and the map

$$t \mapsto \langle \mathcal{B}_0(w + tv)v, v \rangle_V$$

be continuous on $[0, 1]$. Then $\mathcal{B}$ is strongly monotone if and only if $\mathcal{B}_0(w) : V \to V^*$ is strongly monotone, i.e. there is a $\mu_0 > 0$ such that

$$\langle \mathcal{B}_0(w)v, v \rangle_V \geq \mu_0 \|v\|_V^2 \quad \text{for all } w, v \in V$$

PROOF:
Since $\mathcal{B}$ is Fréchet differentiable (with Fréchet derivative $\mathcal{B}_0$) $\mathcal{B}$ is also Gâteaux differentiable, i.e. there exists $\mathcal{B}' : V \to L(V, V^*)$ such that for all $u, v, h \in V$

$$\lim_{t \to 0} \frac{1}{t} \langle \mathcal{B}(u + th)\mathcal{B}(u), v \rangle_V = \langle \mathcal{B}'(u)h, v \rangle_V$$

is satisfied, cf. [GGZ74, Kapitel III, Definition 1.6]. $\mathcal{B}_0$ and $\mathcal{B}'$ coincide. Then the proof from [GGZ74, Kapitel III, Lemma 1.1] can be adopted. Let $\mathcal{B}$ be strongly monotone. For $s \in (0, 1)$ we apply the mean value theorem for a suitable $s_0 \in [0, s]$ and obtain

$$\begin{aligned}
\mu \|sv\|_V^2 &\leq \langle \mathcal{B}(w + sv) - \mathcal{B}(w), sv \rangle_V \\
&= \int_0^s \langle \mathcal{B}_0(w + tv)v, sv \rangle_V \mathrm{d}t \\
&= s^2 \langle \mathcal{B}_0(w + s_0 v)v, v \rangle_V.
\end{aligned}$$

Dividing by s^2 and taking the limit $s \to \infty$ reveals that

$$\langle \mathcal{B}_0(w)v, v \rangle_V \geq \mu_0 \|v\|_V^2$$

with $\mu_0 = \mu$. Conversely, if $\mathcal{B}_0(w)$ is strongly monotone, we have

$$\langle \mathcal{B}(u) - \mathcal{B}(v), u - v \rangle_V = \int_0^1 \langle \mathcal{B}_0(v + t(u-v))(u-v), u-v \rangle_V \mathrm{d}t \geq \mu \|u-v\|_V^2.$$

$\square$

The following Lemma 4.28 and Theorem 4.29 were developed in [MT12] and can be found in [LMT13] in the context for linear ADAEs. The corresponding linear version of Theorem 4.29 was originally proven in [Tis04, Theorem 4.24] in a similar way.

Lemma 4.28.
Let W be a Banach space and let $Z \subseteq H \subseteq Z^*$ be an evolution triple. Let $\mathcal{A}, \mathcal{D} \in L(W, Z)$ be bijective. Define $\mathcal{G} : W \to W^*$ as follows:

$$\langle \mathcal{G}\widetilde{w}, w \rangle_W := (\mathcal{D}\widetilde{w} | \mathcal{A}w)_H \quad \forall \widetilde{w}, w \in W.$$

Then $\mathcal{G} \in L(W, W^*)$ and $\overline{\mathrm{im}\,\mathcal{G}} = W^*$.

PROOF:
The linearity of $\mathcal{G}$ is clear and there is a $\bar{c} > 0$ such that

$$\langle \mathcal{G}\widetilde{w}, w \rangle_W = (\mathcal{D}\widetilde{w} | \mathcal{A}w)_H \leq \bar{c} \|\widetilde{w}\|_W \|w\|_W$$

because the embedding $Z \subseteq H$ is continuous and $\mathcal{A}, \mathcal{D} \in L(W, Z)$. We have to show that for all $\bar{w} \in W^*$ there is a sequence $(\bar{w}_n) \subseteq \mathrm{im}\,\mathcal{G}$ such that

$$\|\bar{w} - \bar{w}_n\|_{W^*} \to 0 \qquad \text{as} \quad n \to \infty.$$

The dual operator $\mathcal{A}^* \in L(Z^*, W^*)$ of $\mathcal{A}$ is bijective. This can be seen as follows. First let $\mathcal{A}^*\bar{z} = 0$ and so $\langle \bar{z}, \mathcal{A}w \rangle_Z = 0$ for all $w \in W$. Since $\mathcal{A}$ is bijective we have $\langle \bar{z}, z \rangle_Z = 0$ for all $z \in Z$ and thus $\bar{z} = 0$. For surjectivity let be $\bar{w} \in W^*$ and define $\bar{z} \in Z^*$ as $\langle \bar{z}, z \rangle_Z := \langle \bar{w}, \mathcal{A}^{-1}z \rangle_W$ because $\mathcal{A}$ is bijective. Then $\mathcal{A}^*\bar{z} = \bar{w}$.
For a given $\bar{w} \in W^*$ the bijectivity of $\mathcal{A}^*$ implies that there is a unique $\bar{z} \in Z^*$ such that $\mathcal{A}^*\bar{z} = \bar{w}$. Since $H^* \subseteq Z^*$ dense there is a sequence $(\bar{u}_n) \subseteq H^*$ such that

$$\forall \varepsilon > 0 \; \exists N \in \mathbb{N} \; \forall n \geq N : \|\bar{z} - \bar{u}_n\|_{Z^*} \leq \frac{\varepsilon}{2}.$$

By the Riesz Representation Theorem A.5 there is a unique $u_n \in H$ such that

$$\langle \bar{u}_n, u \rangle_H = (u_n | u)_H \quad \forall u \in H.$$

Since also $Z \subseteq H$ dense there exists $z_n \in Z$ such that

$$\|z_n - u_n\|_H \leq \frac{\varepsilon}{2c}$$

where $c > 0$ is the constant from the continuous embedding $Z \subseteq H$. We set now

$$\langle \bar{z}_n, z \rangle_Z := (z_n \,|\, z)_H \quad \forall z \in Z.$$

It is $\bar{z}_n \in Z^*$ because the embedding $Z \subseteq H$ is continuous and we have for $z \in Z$

$$
\begin{aligned}
\langle \bar{z} - \bar{z}_n, z \rangle_Z &= \langle \bar{z} - \bar{u}_n, z \rangle_Z + \langle u_n, z \rangle_Z - (z_n \,|\, z)_H \\
&= \langle \bar{z} - \bar{u}_n, z \rangle_Z + (u_n - z_n \,|\, z)_H \\
&\leq \|\bar{z} - \bar{u}_n\|_{Z^*} \|z\|_Z + c \|u_n - z_n\|_H \|z\|_Z \leq \varepsilon \|z\|_Z.
\end{aligned}
$$

We conclude that $\|\bar{z} - \bar{z}_n\|_{Z^*} \to 0$ as $n \to \infty$. Since $\mathcal{D}$ is bijective there is a unique $w_n \in W$ such that $\mathcal{D}w_n = z_n$. We define

$$\langle \bar{w}_n, w \rangle_W := \langle \bar{z}_n, \mathcal{A}w \rangle_Z \quad \forall w \in W.$$

It is $\bar{w}_n \in W^*$ and also $\bar{w}_n \in \operatorname{im} \mathcal{G}$ because

$$\langle \bar{w}_n, w \rangle_W = \langle \bar{z}_n, \mathcal{A}w \rangle_Z = (\mathcal{D}w_n \,|\, \mathcal{A}w)_H = \langle \mathcal{G}w_n, w \rangle_W.$$

Finally, we see for all $w \in W$:

$$
\begin{aligned}
\langle \bar{w} - \bar{w}_n, w \rangle_W &= \langle \mathcal{A}^* \bar{z}, w \rangle_W - \langle \bar{z}_n, \mathcal{A}w \rangle_Z \\
&= \langle \bar{z}, \mathcal{A}w \rangle_Z - \langle \bar{z}_n, \mathcal{A}w \rangle_Z \\
&\leq \tilde{c} \|\bar{z} - \bar{z}_n\|_{Z^*} \|w\|_W
\end{aligned}
$$

for a $\tilde{c} > 0$ because $\mathcal{A}$ is continuous. This completes the proof. $\qquad \square$

Theorem 4.29 (Index characterization).
Let Assumptions 4.1, 4.3, 4.4, 4.7, 4.14 and 4.17 be fulfilled. Let $\mathcal{B}(t) : V \to V^*$ be Fréchet differentiable on V with derivative $\mathcal{B}_0(w,t) : V \to V^*$ and let the map

$$s \mapsto \langle \mathcal{B}_0(w + sv, t)v, v \rangle_V$$

be continuous on $[0,1]$ for all $w, v \in V$, $t \in \mathcal{I}$. Define the operators

$$
\begin{aligned}
\mathcal{G}_0 : V \to V^*, \quad & \langle \mathcal{G}_0 u, v \rangle_V := (\mathcal{D}u \,|\, \mathcal{A}v)_H \\
\mathcal{G}_1(w,t) : V \to V^*, \quad & \langle \mathcal{G}_1(w,t)u, v \rangle_V := (\mathcal{D}u \,|\, \mathcal{A}v)_H + \langle \mathcal{B}_0(w,t)\mathcal{Q}u, v \rangle_V
\end{aligned}
$$

for all $u, v, w \in V$, $t \in \mathcal{I}$. Then the following holds:

(i) $\mathcal{G}_0$ is injective and $\overline{\operatorname{im} \mathcal{G}_0} = V^*$ if and only if $\mathcal{D}$ is injective.

(ii) $\mathcal{G}_1(w,t)$ is injective and $\overline{\operatorname{im} \mathcal{G}_1} = V^*$ for all $w \in V$, $t \in \mathcal{I}$.

PROOF:

(i) If $\mathcal{D}$ is injective, then $\mathcal{D}$ is bijective. If $\mathcal{G}_0 u = 0$ then

$$0 = \langle \mathcal{G}_0 u, u \rangle_V = \|\mathcal{S}\mathcal{D}u\|_H^2 \geq c\|u\|_H^2$$

because of Lemma 4.8 (iii) and the bijectivity of $\mathcal{D}$. So $u = 0$ and $\mathcal{G}_0$ is injective. With Lemma 4.28 we see $\overline{\operatorname{im} \mathcal{G}_0} = V^*$. Conversely, if $\mathcal{D}u = 0$ then $\langle \mathcal{G}_0 u, v \rangle_V = 0$ for all $v \in V$ and so $u = 0$ because $\mathcal{G}_0$ is injective.

(ii) Fix $w \in V$, $t \in \mathcal{I}$ and write $\mathcal{G}_1$ and $\mathcal{B}_0$ instead of $\mathcal{G}_1(w,t)$ and $\mathcal{B}_0(w,t)$, respectively. We show injectivity of $\mathcal{G}_1$ first. Let $u \in \ker \mathcal{G}_1$. This implies

$$(\mathcal{D}u \,|\, Av)_H + \langle \mathcal{B}_0 \mathcal{Q}u, v \rangle_V = 0 \quad \forall v \in V.$$

For $v = \mathcal{Q}u$ we observe

$$\langle \mathcal{B}_0 \mathcal{Q}u, \mathcal{Q}u \rangle_V = 0$$

and thus $\mathcal{Q}u = 0$ because $\mathcal{B}_0$ is strongly monotone, cf. Lemma 4.27. This yields

$$(\mathcal{D}u \,|\, Au)_H = \|\mathcal{S}\mathcal{D}u\|_H^2 = 0$$

if we use $v = u$. Hence $\mathcal{D}u = 0$ because of Lemma 4.8 and so $u \in \operatorname{im} \mathcal{Q}$. Finally, this gives $u = \mathcal{Q}u = 0$.

Next, we have to show that $\overline{\operatorname{im} \mathcal{G}_1} = V^*$, i.e. that for all $\bar{v} \in V^*$ there is a sequence $(\bar{v}_n) \subseteq \operatorname{im} \mathcal{G}_1$ such that

$$\|\bar{v} - \bar{v}_n\|_{V^*} \to 0 \quad \text{as} \quad n \to \infty.$$

Let be $\bar{v} \in V^*$. $U := \operatorname{im} \mathcal{Q}$ is a closed subspace of V because $\mathcal{Q}$ is a projection operator. Hence U is a reflexive Banach space with the norm $\|\cdot\|_V$. We define the operator

$$\widetilde{\mathcal{B}} : U \to U^*, \qquad \langle \widetilde{\mathcal{B}}\tilde{u}, u \rangle_U := \langle \mathcal{B}_0 \mathcal{Q}\tilde{u}, \mathcal{Q}u \rangle_V \quad \forall \tilde{u}, u \in U$$

and $\widetilde{\mathcal{B}}$ is well-defined. It is strongly monotone and bounded because $\mathcal{B}_0$ is and the norm on U is $\|\cdot\|_V$. Define furthermore

$$\langle \bar{u}, u \rangle_U := \langle \bar{v}, \mathcal{Q}u \rangle_V \quad \forall u \in U$$

and clearly $\bar{u} \in U^*$ because $\bar{v} \in V^*$. With the Browder-Minty Theorem A.14 we can solve uniquely the equation $\widetilde{\mathcal{B}}\tilde{u} = \bar{u}$ or equivalently

$$\langle \widetilde{\mathcal{B}}\tilde{u}, u \rangle_U = \langle \bar{v}, \mathcal{Q}u \rangle_V \tag{4.33}$$

and get the solution $\tilde{u} \in U$. Let $\mathcal{P}$ be the complementary projection operator of $\mathcal{Q}$ and set $W := \operatorname{im} \mathcal{P}$. W is a Banach space with the norm $\|\cdot\|_V$. Furthermore we set

$$\langle \bar{w}, w \rangle_W := \langle \bar{v} - \mathcal{B}_0 \mathcal{Q}\tilde{u}, \mathcal{P}w \rangle_V \quad \forall w \in W \tag{4.34}$$

Clearly $\bar{w} \in W^*$ because $\bar{v} \in V^*$ and $\mathcal{B}_0$ is bounded. The linear and continuous maps

$$\mathcal{D}|_W, \ \mathcal{A}|_W : W \to Z,$$

are bijective because $\mathcal{D}$ and $\mathcal{A} = \mathcal{T}\mathcal{D}$ are surjective and $\ker \mathcal{D} = \operatorname{im} \mathcal{Q} = \ker \mathcal{P}$. Setting $\mathcal{G} : W \to W^*$ as

$$\langle \mathcal{G}\widetilde{w}, w \rangle_W := (\mathcal{D}|_W \widetilde{w} \,|\, \mathcal{A}|_W w)_H = (\mathcal{D}\widetilde{w} \,|\, \mathcal{A}w)_H \quad \forall \widetilde{w}, w \in W$$

we can apply Lemma 4.28. So there is a sequence $(\bar{w}_n) \subseteq \operatorname{im} \mathcal{G}$ such that

$$\|\bar{w} - \bar{w}_n\|_{W^*} \to 0 \qquad \text{as} \quad n \to \infty.$$

$\bar{w}_n \in \operatorname{im} \mathcal{G}$ means that there exists a $\widetilde{w}_n \in W$ such that

$$(\mathcal{D}\widetilde{w}_n \,|\, \mathcal{A}w)_H = \langle \mathcal{G}\widetilde{w}_n, w \rangle_W = \langle \bar{w}_n, w \rangle_W \quad \forall w \in W. \tag{4.35}$$

We define now

$$\widetilde{v}_n := \mathcal{P}\widetilde{w}_n + \mathcal{Q}\widetilde{u} \in V \tag{4.36}$$

and set

$$\langle \bar{v}_n, v \rangle_V := \langle \mathcal{G}_1 \widetilde{v}_n, v \rangle_V \quad \forall v \in V.$$

By definition $\bar{v}_n \in \operatorname{im} \mathcal{G}_1$ and we observe for all $v \in V$:

$$
\begin{aligned}
\langle \bar{v} - \bar{v}_n, v \rangle_V \ &= \ \langle \bar{v}, v \rangle_V - \langle \mathcal{G}_1 \widetilde{v}_n, v \rangle_V \\
&\overset{(4.36)}{=} \ \langle \bar{v}, v \rangle_V - (\mathcal{D}\widetilde{w}_n \,|\, \mathcal{A}v)_H - \langle \mathcal{B}_0 \mathcal{Q}\widetilde{u}, v \rangle_V \\
&= \ \langle \bar{v} - \mathcal{B}_0 \mathcal{Q}\widetilde{u}, \mathcal{P}v \rangle_V + \langle \bar{v} - \mathcal{B}_0 \mathcal{Q}\widetilde{u}, \mathcal{Q}v \rangle_V - (\mathcal{D}\widetilde{w}_n \,|\, \mathcal{A}v)_H \\
&\overset{(4.33),(4.34)}{=} \ \langle \bar{w}, \mathcal{P}v \rangle_W - (\mathcal{D}\widetilde{w}_n \,|\, \mathcal{A}v)_H \\
&\overset{(4.35)}{=} \ \langle \bar{w} - \bar{w}_n, \mathcal{P}v \rangle_W \\
&\leq \ \widetilde{c} \|\bar{w} - \bar{w}_n\|_{W^*} \|v\|_V.
\end{aligned}
$$

So we obtain $\|\bar{v} - \bar{v}_n\|_{V^*} \to 0$ as $n \to \infty$. $\qquad\qquad\square$

Remark 4.30 (ADAE Index of (4.4)).
If we identify $\mathcal{D}u \in Z$ with the unique element in Z^* satisfying

$$\langle \mathcal{D}u, z \rangle_Z = (\mathcal{D}u \,|\, z)_H \quad \forall z \in Z$$

we can write

$$\mathcal{G}_0 = \mathcal{A}^* \mathcal{D}, \quad \mathcal{G}_1(w, t) = \mathcal{A}^* \mathcal{D} + \mathcal{B}_0(w, t) \mathcal{Q}_0$$

which matches the definition of $\mathcal{G}_0$ and $\mathcal{G}_1$ in Definition 2.3. In this sense with Theorem 4.29 and V, Z being Hilbert spaces the ADAE (4.4) has at most ADAE Index 1 and it has ADAE Index 0 if and only if $\mathcal{D}$ is injective.

4.5. Conclusion

In this chapter we have studied abstract differential-algebraic equations (ADAEs) of the form (4.4) with regard to unique solvability, convergence of Galerkin solutions, perturbation results and the ADAE Index. The results presented in this chapter are generalizations of the linear case in [Tis04] where the operator $\mathcal{B}(t)$ is linear and the structural condition $\mathcal{A} = \mathcal{D}$ holds. We assume the possibly nonlinear operator $\mathcal{B}(t)$ to be strongly monotone and to be bounded by a growth condition (Assumption 4.14). The structural assumption in the dynamical part of (4.4) is generalized by requiring $\mathcal{A} = \mathcal{T}\mathcal{D}$ for a suitable operator $\mathcal{T}$ (Assumption 4.7). The requirements for $\mathcal{T}$ are such that the crucial integration by parts formula can still be applied. The form (4.4) includes the corresponding well-known case for evolution equations, cf. [Zei90b, Theorem 30.A]. However, the choice of the basis functions is not completely arbitrary because it needs to be ensured that the Galerkin solution satisfies the inherent constraints, cf. Assumption 4.17 in combination with Theorem 4.24.

The unique solvability result of (4.4) in Theorem 4.24 is based on a priori estimates of the Galerkin solutions and then showing the weak convergence of these to a solution of (4.4). For the definition of the derivative $[\mathcal{D}u]'$ a splitting condition concerning the nullspace of $\mathcal{D}$ is needed, cf. Assumption 4.4. No standard solvability results from the theory of differential-algebraic equations can be applied to solve the Galerkin equations (4.7) because the right hand side of (4.4) is only assumed to be L_2-integrable in time (Remark 4.23). The unique solvability of the Galerkin equations is shown in Theorem 4.22. It is based on first rewriting them in a semi-explicit form (4.13) and then applying the Theorem of Carathéodory, cf. Theorem A.3. It is also important that we achieved a global solution for every Galerkin step n on the given interval $\mathcal{I}$. Furthermore strong convergence of the Galerkin solutions is shown in Theorem 4.25.

In addition we analyzed the ADAE (4.4) in the sense of the Perturbation Index and the ADAE Index from chapter 2. We showed that (4.4) has Perturbation Index 1 whereas the definition of the ADAE Index could not be applied directly because it relies on having smoother solutions. However, we defined suitable operators $\mathcal{G}_0$ and $\mathcal{G}_1$ showing that (4.4) in this sense has at most ADAE Index 1.

The results in this chapter were restricted to L_2-spaces in time, but it seems to be possible to extend the solvability and the perturbation result to L_p-theory. More precisely, let $p \geq 2$, $p^{-1} + q^{-1} = 1$. With $r \in L_q(\mathcal{I}, V^*)$ we would obtain a solution u in the space

$$W^1_{p,\mathcal{D}} := \left\{ u \in L_p(\mathcal{I}, V) |\ [\mathcal{D}u]' \in L_q(\mathcal{I}, Z^*) \right\}.$$

The requirements for the operator family $\mathcal{B}(t)$ have to be adjusted as well. We refer to the L_p-theory for nonlinear evolution equations in e.g. [Zei90b, Emm04, Rou05].

Furthermore we have seen that the integration by parts formula is as important for the solvability result as it is for ordinary evolution equations. We put strong conditions on the operators $\mathcal{A}, \mathcal{D}$ and $\mathcal{T}$ to be able to use it. However, having coupled systems in mind nonlinearities in the dynamical part would also be desirable. In circuit simulation when systems of partial differential equations are coupled to the equations of the Modified Nodal Analysis they occur in the finite dimensional part of the coupled system, see chapter 3. Additionally, because the entire operator $\mathcal{B}(t)$ cannot be expected to be strongly monotone in general, we will investigate prototype coupled systems in the next chapter.

5. Coupled Systems

As outlined in chapter 2 coupled systems are of great importance in many application areas. Especially in circuit simulation many examples can be found. Basic circuit elements are described by constitutive relations within the framework of the Modified Nodal Analysis (MNA), but more complex elements like diodes or transistors can be described by equivalent circuits which are modeled by a set of basic elements. Due to miniaturization in chip design, higher package densities and higher operating frequencies the standard equations of the MNA and these equivalent circuits are not sufficient anymore for simulating correct physical behavior. To capture the behavior of distinguished elements or certain effects involved systems of partial differential equations (PDEs) are added to the circuit description. The drift diffusion equations for simulating semiconductors, cf. [ABGT03, ABG05, ABG10], the heat equation for simulating heating effects, cf. [Bar04, Cul09], or the Maxwell equations for simulating electromagnetic behavior of certain devices, cf. [Sch11, Bau12], for example, have been added. These systems interchange information via certain coupling terms, additional coupling equations, certain variables serving as input for boundary conditions of the PDE system or even more general parametric coupling. This results in the overall system being very complex and these specific coupled systems themselves have already become research topics.

In general these coupled systems are treated numerically by approximating the system in space first. Finite elements, for example, can be used for semiconductors, cf. [ST05], or heating effects, cf. [Cul09], and the finite integration technique is an established discretization scheme for electromagnetic devices, cf. [Bau12]. The resulting semi-discretized system is a differential-algebraic equation and the numerical analysis is mainly based on this system, concerning index analysis, sensitivity with regard to perturbations or stability and consistency of numerical methods for time integration. However, the difference of the semi-discretized system to the original coupled system has merely been analyzed. For a coupled system involving stationary semiconductor behavior a convergence estimate for the whole coupled system was proven in [MT11]. Furthermore it is important to investigate the sensitivity of the coupled system itself with regard to perturbations, cf. e.g. [Bod07]. It is known that different discretization schemes may act as a regularization (decreasing the index) or even a deregularization (increasing the index) of the coupled system, cf. [Gün00].

In this chapter we will present two prototype coupled systems as a starting point for treating (nonlinear) coupled systems systematically. For both systems we give unique solvability results, investigate the convergence of Galerkin solutions and present a perturbation estimate. These prototype systems are also applicable to circuit simulation as we will demonstrate on two exemplary systems. Relevant background material for this chapter can be found in Appendix A.3, A.4 and A.5.

5.1. An elliptic prototype

In this section we discuss an elliptic prototype of a coupled system. Let $\mathcal{I} := [t_0, T]$ be an interval and V be a real Banach space. Consider the system

$$\frac{\mathrm{d}}{\mathrm{d}t} m(x(t), t) + f(z(t), u(t), t) = 0, \quad t \in \mathcal{I}, \tag{5.1a}$$

$$g(z(t), t) = 0, \tag{5.1b}$$

$$\mathcal{B}(u(t)) + \mathcal{R}(z(t), t) = 0, \quad \text{in } V^*, \tag{5.1c}$$

$$x(t_0) = x_0 \in \mathbb{R}^{n_x} \tag{5.1d}$$

with functions $m : \mathbb{R}^{n_x} \times \mathcal{I} \to \mathbb{R}^{n_x}$, $f : \mathbb{R}^{n_z} \times V \times \mathcal{I} \to \mathbb{R}^{n_x}$, $g : \mathbb{R}^{n_z} \times \mathcal{I} \to \mathbb{R}^{n_y}$ and operators $\mathcal{B} : V \to V^*$ and $\mathcal{R} : \mathbb{R}^{n_z} \times \mathcal{I} \to V^*$. The unknowns are $x(t) \in \mathbb{R}^{n_x}$, $y(t) \in \mathbb{R}^{n_y}$ and $u(t) \in V$ for $t \in \mathcal{I}$. We also use the convention to write $z(t) = (x(t), y(t))^\top \in \mathbb{R}^{n_z}$, $n_z = n_x + n_y$. We will also often omit the explicit time dependency of the variables. The initial value $x_0 \in \mathbb{R}^{n_x}$ is given. Note that equations (5.1a), (5.1b) represent a semi-linear (finite dimensional) DAE whereas (5.1c) is an (infinite dimensional) operator equation. The coupling of these two systems is realized by letting f depend on u and $\mathcal{R}$ depend on z. We will investigate system (5.1) w.r.t. solvability, perturbation estimates and Galerkin convergence under suitable assumptions. We first prove unique solvability.

Theorem 5.1 (Unique solvability).
Let $\mathcal{I} := [t_0, T]$ be an interval and assume the following:

(i) V is a real reflexive Banach space.

(ii) $m \in C^1(\mathbb{R}^{n_x} \times \mathcal{I}, \mathbb{R}^{n_x})$ is strongly monotone w.r.t. x.

(iii) $f \in C(\mathbb{R}^{n_z} \times V \times \mathcal{I}, \mathbb{R}^{n_x})$ is Lipschitz continuous w.r.t. z and u.

(iv) $g \in C(\mathbb{R}^{n_z} \times \mathcal{I}, \mathbb{R}^{n_y})$ is uniquely solvable w.r.t. $y \in \mathbb{R}^{n_y}$, i.e. there is a solution function $\psi_g \in C(\mathbb{R}^{n_x} \times \mathcal{I}, \mathbb{R}^{n_y})$ such that $y = \psi_g(x, t)$ whenever $g((x, y)^\top, t) = 0$ for all x, y, t. Furthermore ψ_g is Lipschitz continuous w.r.t. x.

(v) $\mathcal{B} : V \to V^*$ is strongly monotone and hemicontinuous.

(vi) $\mathcal{R} \in C(\mathbb{R}^{n_z} \times \mathcal{I}, V^*)$ is Lipschitz continuous w.r.t. z.

Then (5.1) has a unique solution $(z, u) \in C(\mathcal{I}, \mathbb{R}^{n_z} \times V)$ with $x \in C^1(\mathcal{I}, \mathbb{R}^{n_x})$ for every initial value $x_0 \in \mathbb{R}^{n_x}$.

PROOF:

Existence. For any given $z \in \mathbb{R}^{n_z}$, $t \in \mathcal{I}$ we can solve the operator equation (5.1c) uniquely due to the Browder-Minty Theorem A.14. We obtain

$$u = \mathcal{B}^{-1}(-\mathcal{R}(z,t))$$

and insert it into (5.1a). We remark that $\mathcal{B}^{-1} : V^* \to V$ is Lipschitz continuous, so u depends continuously on (z,t) and is Lipschitz continuous w.r.t. z. Equation (5.1a) now reads

$$\frac{\mathrm{d}}{\mathrm{d}t} m(x,t) = -f((x,y)^\top, \mathcal{B}^{-1}(-\mathcal{R}((x,y)^\top,t)),t) =: \widetilde{f}(x,y,t)$$

with $\widetilde{f} : \mathbb{R}^{n_x} \times \mathbb{R}^{n_y} \times \mathcal{I} \to \mathbb{R}^{n_x}$ being continuous and Lipschitz continuous w.r.t. x and y. We can now apply Theorem 3.15 (in combination with Remark 3.18) to get a unique solution $z_* = (x_*, y_*)$ of the system

$$\frac{\mathrm{d}}{\mathrm{d}t} m(x,t) = \widetilde{f}(x,y,t), \quad t \in \mathcal{I}$$
$$0 = g(x,y,t)$$

with the initial value $x_*(t_0) = x_0$. Here z_* is continuous and x_* is continuously differentiable. So setting

$$u_*(t) := \mathcal{B}^{-1}(-\mathcal{R}(z_*(t),t)), \quad t \in \mathcal{I}$$

we see that $u_* \in C(\mathcal{I}, V)$ and that (z_*, u_*) is a solution to (5.1).

Uniqueness. Let there be two solutions (z_i, u_i), $i = 1, 2$, of (5.1). The strong monotonicity of $\mathcal{B}$ gives

$$\mu \|u_1(t) - u_2(t)\|_V^2 \quad \leq \quad \langle \mathcal{B}(u_1(t)) - \mathcal{B}(u_2(t)), u_1(t) - u_2(t) \rangle_V$$
$$\overset{(5.1c)}{=} \quad \langle \mathcal{R}(z_2(t),t) - \mathcal{R}(z_1(t),t), u_1(t) - u_2(t) \rangle_V$$

for a $\mu > 0$ and all $t \in \mathcal{I}$ and hence

$$\|u_1(t) - u_2(t)\|_V \leq \frac{1}{\mu} \|\mathcal{R}(z_2(t),t) - \mathcal{R}(z_1(t),t)\|_{V^*}.$$

Furthermore we obtain from (5.1b) using the solution function ψ_g that

$$\|y_1(t) - y_2(t)\| = \|\psi_g(x_1(t),t) - \psi_g(x_2(t),t)\| \leq L_{\psi_g} \|x_1(t) - x_2(t)\|$$

for a $L_{\psi_g} > 0$. Thus there is a constant $c_u > 0$ such that

$$\|u_1(t) - u_2(t)\|_V \leq c_u \|x_1(t) - x_2(t)\|.$$

because $\mathcal{R}$ is Lipschitz continuous w.r.t. z and

$$\|z_1(t) - z_2(t)\| \le \|x_1(t) - x_2(t)\| + \|y_1(t) - y_2(t)\|.$$

Now integrating (5.1a) over $[t_0, t]$ gives

$$m(x_1(t), t) - m(x_2(t), t) = \int_{t_0}^{t} f(z_2(s), u_2(s), s) - f(z_1(s), u_1(s), s)\mathrm{d}s.$$

Using now the strong monotonicity of m, the Lipschitz continuity of f and the Gronwall Lemma in combination with the estimates above we see that $x_1 = x_2$. Compare the proof of Theorem 3.15 for more details. Hence we also get $y_1 = y_2$ and $u_1 = u_2$. $\square$

Note that the assumptions on the finite dimensional part (5.1a), (5.1b) are very similar to the assumptions required in Theorem 3.15. Especially sufficient conditions for (iv) are discussed in Remark 3.8. We now investigate the solution to (5.1) w.r.t. perturbations.

Theorem 5.2 (Perturbation estimate).
Let $\mathcal{I} := [t_0, T]$ be an interval. Consider the system

$$\frac{\mathrm{d}}{\mathrm{d}t}m(x(t), t) + f(z(t), u(t), t, \delta_x(t)) = 0, \quad t \in \mathcal{I}, \tag{5.2a}$$

$$g(z(t), t, \delta_y(t)) = 0, \tag{5.2b}$$

$$\mathcal{B}(u(t)) + \mathcal{R}(z(t), t) + \delta_u(t) = 0, \quad in\ V^*, \tag{5.2c}$$

$$x(t_0) = x_0^\delta \in \mathbb{R}^{n_x} \tag{5.2d}$$

and assume the following:

(i) V *is a real reflexive Banach space.*

(ii) $m \in C^1(\mathbb{R}^{n_x} \times \mathcal{I}, \mathbb{R}^{n_x})$ *is strongly monotone w.r.t. x.*

(iii) $f \in C(\mathbb{R}^{n_z} \times V \times \mathcal{I} \times \mathbb{R}^{n_x}, \mathbb{R}^{n_x})$ *is Lipschitz continuous w.r.t. z, u and $\delta_x(t)$.*

(iv) $g \in C(\mathbb{R}^{n_z} \times \mathcal{I} \times \mathbb{R}^{n_y}, \mathbb{R}^{n_y})$ *is uniquely solvable w.r.t. $y \in \mathbb{R}^{n_y}$, i.e. there is a solution function $\psi_g^\delta \in C(\mathbb{R}^{n_x} \times \mathcal{I} \times \mathbb{R}^{n_y}, \mathbb{R}^{n_y})$ such that $y = \psi_g^\delta(x, t, \delta_y(t))$ whenever $g((x, y)^\top, t, \delta_y(t)) = 0$ for all x, y, t. Furthermore ψ_g^δ is Lipschitz continuous w.r.t. x and $\delta_y(t)$.*

(v) $\mathcal{B} : V \to V^*$ *is strongly monotone and hemicontinuous.*

(vi) $\mathcal{R} : \mathbb{R}^{n_z} \times \mathcal{I} \to V^*$ *is continuous and Lipschitz continuous w.r.t. z.*

(vii) $\delta_x \in C(\mathcal{I}, \mathbb{R}^{n_x})$, $\delta_y \in C(\mathcal{I}, \mathbb{R}^{n_y})$, $\delta_u \in C(\mathcal{I}, V^*)$.

Then (5.2) has a unique solution $(z^\delta, u^\delta) \in C(\mathcal{I}, \mathbb{R}^{n_z} \times V)$ with $x^\delta \in C^1(\mathcal{I}, \mathbb{R}^{n_z})$ for every initial value $x_0^\delta \in \mathbb{R}^{n_z}$. Let (z, u) be the solution for $(\delta_x, \delta_y, \delta_u) = 0$ with initial value $x(t_0) = x_0$. If $\|x_0 - x_0^\delta\|$ is sufficiently small then there is a $C > 0$ such that

$$\left\|z - z^\delta\right\|_\infty + \max_{t \in \mathcal{I}} \left\|u(t) - u^\delta(t)\right\|_V \leq C \left(\left\|x_0 - x_0^\delta\right\| + \left\|\delta_x\right\|_\infty + \left\|\delta_y\right\|_\infty + \max_{t \in \mathcal{I}} \left\|\delta_u(t)\right\|_{V^*} \right).$$

PROOF:

Solvability. For given perturbations δ_x, δ_y and δ_u we set

$$f^\delta(z, u, t) := f(z, u, t, \delta_x(t)),$$
$$g^\delta(z, t) := g(z, t, \delta_y(t)),$$
$$\mathcal{R}^\delta(z, t) := \mathcal{R}(z, t) + \delta_u(t)$$

for all $t \in \mathcal{I}$ and $z \in \mathbb{R}^{n_z}$, $u \in V$. Then the functions f^δ, g^δ and $\mathcal{R}^\delta$ inherit all the properties from the functions f, g and $\mathcal{R}$ and Theorem 5.1 is applicable. Hence we get the desired unique solution (z^δ, u^δ) for the initial value x_0^δ. In the case of $(\delta_x, \delta_y, \delta_u) = 0$ and given initial value x_0 we denote the solution by (z, u).

Perturbation estimate. Using the strong monotonicity of $\mathcal{B}$ gives

$$
\begin{aligned}
\left\|u(t) - u^\delta(t)\right\|_V \quad &\leq \quad \frac{1}{\mu} \left\|\mathcal{B}(u(t)) - \mathcal{B}(u^\delta(t))\right\|_{V^*} \\
&\overset{(5.2c)}{\leq} \quad \frac{1}{\mu} \left\|\mathcal{R}(z(t), t) - \mathcal{R}(z^\delta(t), t)\right\|_{V^*} + \frac{1}{\mu} \left\|\delta_u(t)\right\|_{V^*} \\
&\leq \quad c_u \left(\left\|x(t) - x^\delta(t)\right\| + \left\|y(t) - y^\delta(t)\right\| + \left\|\delta_u(t)\right\|_{V^*} \right)
\end{aligned}
$$

for constants $\mu, c_u > 0$. In the last line we used the Lipschitz continuity of $\mathcal{R}$ w.r.t. z. Furthermore we have with (5.2b) that

$$y^\delta(t) = \psi_g(x^\delta(t), t, \delta_y(t)), \quad y(t) = \psi_g(x(t), t, 0), \quad t \in \mathcal{I}$$

and we obtain

$$
\begin{aligned}
\left\|y(t) - y^\delta(t)\right\| &\leq \left\|\psi_g(x(t), t, 0) - \psi_g(x^\delta(t), t, 0)\right\| + \left\|\psi_g(x^\delta(t), t, 0) - \psi_g(x^\delta(t), t, \delta_y(t))\right\| \\
&\leq c_y \left(\left\|x(t) - x^\delta(t)\right\| + \left\|\delta_y(t)\right\| \right)
\end{aligned}
$$

for a constant $c_y > 0$. Applying the same arguments as in the proof of Theorem 3.17 (in combination with Remark 3.18) the desired estimate follows. The main steps in the proof of Theorem 3.17 were integrating (5.2b) over $[t_0, t]$, making use of the Lipschitz continuity of f and applying the mean value theorem ($\|x_0 - x_0^\delta\|$ is sufficiently small). □

Theorem 5.2 shows that (5.1) has Perturbation Index 1 under the given assumptions, cf. Definition 2.4. We do not require the perturbations $(\delta_x, \delta_y, \delta_u)$ to be sufficiently small.

This is due to the Lipschitz continuity of the functions f, g and $\mathcal{R}$.

The operator equation (5.1c) is stated for a possibly infinite dimensional Banach space V. So corresponding Galerkin equations can be formulated and we show that the solutions to the Galerkin equations converge uniformly to the solution of the system (5.1).

Theorem 5.3 (Convergence of Galerkin solutions).
Let $\mathcal{I} := [t_0, T]$ be an interval. Consider system (5.1) with initial value $x_0 \in \mathbb{R}^{n_x}$ and assume the following:

(i) *V is a real reflexive and separable Banach space. Let $\dim V = \infty$ and $\{v_1, v_2, \dots\}$ be a basis of V in the sense of Definition A.11. Set $V_n := \{v_1, \dots, v_n\}$.*

(ii) *$m \in C^1(\mathbb{R}^{n_x} \times \mathcal{I}, \mathbb{R}^{n_x})$ is strongly monotone w.r.t. x.*

(iii) *$f \in C(\mathbb{R}^{n_z} \times V \times \mathcal{I}, \mathbb{R}^{n_x})$ is Lipschitz continuous w.r.t. z and u.*

(iv) *$g \in C(\mathbb{R}^{n_z} \times \mathcal{I}, \mathbb{R}^{n_y})$ is uniquely solvable w.r.t. $y \in \mathbb{R}^{n_y}$, i.e. there is a solution function $\psi_g \in C(\mathbb{R}^{n_x} \times \mathcal{I}, \mathbb{R}^{n_y})$ such that $y = \psi_g(x, t)$ whenever $g((x, y)^\top, t) = 0$ for all x, y, t. Furthermore ψ_g is Lipschitz continuous w.r.t. x.*

(v) *$\mathcal{B} : V \to V^*$ is strongly monotone and hemicontinuous.*

(vi) *$\mathcal{R} : \mathbb{R}^{n_z} \times \mathcal{I} \to V^*$ can be decomposed as*

$$\mathcal{R}(z, t) = \widetilde{\mathcal{R}}(z) + r(t) \quad \forall (z, t) \in \mathbb{R}^{n_z} \times \mathcal{I}$$

where $\widetilde{\mathcal{R}} \in C(\mathbb{R}^{n_z}, V^*)$, $r \in C(\mathcal{I}, V^*)$ and $\widetilde{\mathcal{R}}$ is Lipschitz continuous.

Then the Galerkin equations

$$\frac{\mathrm{d}}{\mathrm{d}t} m(x_n(t), t) + f(z_n(t), u_n(t), t) = 0, \quad t \in \mathcal{I}, \tag{5.3a}$$

$$g(z_n(t), t) = 0, \tag{5.3b}$$

$$\langle \mathcal{B}(u_n(t)) + \mathcal{R}(z_n(t), t), v_i \rangle_V = 0, \quad i = 1, \dots, n, \tag{5.3c}$$

$$x_n(t_0) = x_0 \in \mathbb{R}^{n_x} \tag{5.3d}$$

are uniquely solvable with solution $(z_n, u_n) \in C(\mathcal{I}, \mathbb{R}^{n_z} \times V_n)$ and $x_n \in C^1(\mathcal{I}, \mathbb{R}^{n_x})$. Furthermore, if (z, u) is the solution to (5.1), then

$$\|z - z_n\|_\infty + \max_{t \in \mathcal{I}} \|u(t) - u_n(t)\|_V \to 0 \qquad \text{as } n \to \infty.$$

PROOF:
We outline first the strategy of the proof and show the following steps.
Step 1. The Galerkin equations (5.3) are uniquely solvable.

Step 2. For the Galerkin solutions (z_n, u_n) a priori estimates hold. More precisely, there is a $C > 0$ (independent of n and t) such that for all $t \in \mathcal{I}$:

$$\|z_n(t)\| \leq C, \quad \|u_n(t)\|_V \leq C, \quad \|\mathcal{B}(u_n(t))\|_{V^*} \leq C.$$

Step 3. The set $\{u_n | \; n \in \mathbb{N}\}$ is equicontinuous on $\mathcal{I}$.

Step 4. There is a subsequence n' and $(z_*, u_*) \in C(\mathcal{I}, \mathbb{R}^{n_z} \times V)$ such that

$$z_{n'} \rightrightarrows z_*, \quad u_{n'}(t) \to u_*(t) \text{ in } V, \; t \in \mathcal{I} \quad \text{as } n' \to \infty.$$

Step 5. The limits (z_*, u_*) fulfill system (5.1) and

$$z_n \rightrightarrows z_*, \quad u_n \rightrightarrows u_* \quad \text{as } n \to \infty.$$

Ad (1). Equation (5.3c) can be equivalently formulated as

$$\langle \mathcal{B}(u_n(t)) + \mathcal{R}(z_n(t), t), v \rangle_V = 0 \quad \forall v \in V_n \tag{5.4}$$

We identify $V_n \subseteq V$ as a subspace of V with the induced norm and denote the natural embedding operator from V_n to V by $\mathcal{J}_n : V_n \to V$. The dual operator of $\mathcal{J}_n$ is denoted by $\mathcal{J}_n^* : V^* \to V_n^*$. Since we have $\mathcal{J}_n u_n(t) = u_n(t)$ we can write equation (5.4) equivalently in operator notation as follows

$$\mathcal{B}_n(u_n(t)) + \mathcal{R}_n(z_n(t), t) = 0 \quad \text{in } V_n^*$$

with $\mathcal{B}_n := \mathcal{J}_n^* \mathcal{B} \mathcal{J}_n : V_n \to V_n^*$ and $\mathcal{R}_n(z_n(t), t) := \mathcal{J}_n^* \mathcal{R}(z_n(t), t)$, $t \in \mathcal{I}$. With Lemma A.15 $\mathcal{B}_n$ is strongly monotone and hemicontinuous because $\mathcal{B}$ is. The properties of $\mathcal{R}$ transfer to $\mathcal{R}_n$ as well. Hence we can apply Theorem 5.1 and obtain a unique solution $(z_n, u_n) \in C(\mathcal{I}, \mathbb{R}^{n_z} \times V_n) \subseteq C(\mathcal{I}, \mathbb{R}^{n_z} \times V)$ to (5.3) with $x_n \in C^1(\mathcal{I}, \mathbb{R}^{n_x})$.

Ad (2). First we examine the algebraic part (5.3b). With the Lipschitz continuity of ψ_g we get

$$\|y_n(t)\| = \|\psi_g(x_n(t), t) - \psi_g(0, t) + \psi_g(0, t)\| \leq c_y \|x_n(t)\| + d_y$$

and the strong monotonicity of $\mathcal{B}$ ($\mu_\mathcal{B} > 0$) together with the Lipschitz continuity of $\mathcal{R}$ implies

$$\begin{aligned}
\|u_n(t)\|^2 &\leq \frac{1}{\mu_\mathcal{B}} \langle \mathcal{B}(u_n(t)) - \mathcal{B}(0), u_n(t) \rangle_V \\
&\leq \frac{1}{\mu_\mathcal{B}} \left(\|\mathcal{R}(z_n(t))\|_{V^*} + \|\mathcal{B}(0)\|_{V^*} \right) \|u_n(t)\|_V \\
&\leq (c_u \|z_n(t)\| + d_u) \|u_n(t)\|_V
\end{aligned}$$

with constants c_y, c_u, d_y, $d_u > 0$. The relevant requirements here were ψ_g and $\mathcal{R}$ being continuous and $\mathcal{I}$ being compact. We integrate (5.3a) over $[t_0, t]$ and get

$$m(x_n(t), t) = m(x_0, t_0) - \int_{t_0}^{t} f(z_n(s), u_n(s), s) \mathrm{d}s. \tag{5.5}$$

The strong monotonicity of m gives

$$\|x_n(t) - x_0\| \leq \frac{1}{\mu} \|m(x_n(t), t) - m(x_0, t)\|$$

$$\leq \|m(x_0, t_0) - m(x_0, t)\| + \int_{t_0}^{t} \|f(z_n(s), u_n(s), s)\| \, ds$$

$$\leq \|m(x_0, t_0) - m(x_0, t)\| + \int_{t_0}^{t} c_1 (\|z_n(s)\| + \|u_n(s)\|_V) + \|f(0, 0, s)\| \, ds$$

$$\leq d_x + c_x \int_{t_0}^{t} \|x_n(s)\| \, ds$$

with constants c_1, c_x, $d_x > 0$. For the last line we inserted the estimates from before and we used that m and f are continuous. An application of Gronwall's Lemma reveals that $x_n(t)$ is uniformly bounded. After insertion into the estimates before we obtain that $\|z_n(t)\|$ and $\|u_n(t)\|_V$ are uniformly bounded by a constant $\widetilde{C} > 0$. It remains to show that $\mathcal{B}(u_n(t))$ is also bounded. Since $\mathcal{B}$ is monotone it is also locally bounded, i.e. there exist $r, \alpha > 0$ such that

$$\|v\|_V \leq r \quad \Rightarrow \quad \|\mathcal{B}(v)\|_{V^*} \leq \alpha,$$

cf. [Zei90b, Proposition 26.4 (a)]. From (5.3c) we infer

$$|\langle \mathcal{B}(u_n(t)), u_n(t) \rangle_V| = |\langle \mathcal{R}(z_n(t), t), u_n(t) \rangle_V| \leq \|\mathcal{R}(z_n(t), t)\|_{V^*} \|u_n(t)\|_V \leq C_{\mathcal{B}}.$$

for a $C_{\mathcal{B}} > 0$ because $\mathcal{R}$ is continuous and $z_n(t)$, $u_n(t)$ are bounded. Since

$$\langle \mathcal{B}(u_n(t)) - \mathcal{B}(v), u_n(t) - v \rangle \geq 0 \quad \forall v \in V$$

it is

$$\|\mathcal{B}(u_n(t))\|_{V^*} = \sup_{\|v\|=r} r^{-1} \langle \mathcal{B}(u_n(t)), v \rangle_V$$

$$\leq \sup_{\|v\|=r} r^{-1} (\langle \mathcal{B}(v), v \rangle_V + \langle \mathcal{B}(u_n(t)), u_n(t) \rangle_V - \langle \mathcal{B}(v), u_n(t) \rangle_V)$$

$$\leq r^{-1} \left(\alpha r + C_{\mathcal{B}} + \alpha \widetilde{C} \right).$$

The right hand side is independent of t or n.
Ad (3). We show first that $\{x_n | \ n \in \mathbb{N}\} \subseteq C(\mathcal{I}, \mathbb{R}^{n_x})$ is relatively compact. We remark that

$$\|f(z_n(t), u_n(t), t)\| \leq C_f \tag{5.6}$$

for a constant $C_f > 0$ (independent of n and t) because of the a priori estimates and the Lipschitz continuity of f. We define

$$w_n(t) := m(x_n(t), t), \quad n \in \mathbb{N}, \ t \in \mathcal{I}$$

and clearly have $w_n \in C(\mathcal{I}, \mathbb{R}^{n_x})$ because x_n and m are continuous. With (5.5) and (5.6) we see that $\|w_n(t)\| \leq C_w$ with a constant $C_w > 0$ being independent of n and t. Furthermore the w_n are equicontinuous because

$$\|w_n(t) - w_n(\bar{t})\| \overset{(5.5)}{\leq} \int_{\bar{t}}^{t} \|f(z_n(s), u_n(s), s)\| \, \mathrm{d}s \leq C_f \, |t - \bar{t}| \, .$$

An application of the Arzelà-Ascoli-Theorem A.19 gives a subsequence w_k and a limit $w \in C(\mathcal{I}, \mathbb{R}^{n_x})$ with $w_k \rightrightarrows w$ as $k \to \infty$. Define $x \in C(\mathcal{I}, \mathbb{R}^{n_x})$ as the unique solution of $m(x(t), t) = w(t)$, $t \in \mathcal{I}$, cf. Lemma 3.7. Then using the strong monotonicity of m we obtain

$$\begin{aligned} \mu \|x_k(t) - x(t)\| &\leq \|m(x_k(t), t) - m(x(t), t)\| \\ &= \|w_k(t) - w(t)\| \\ &\leq \max_{t \in \mathcal{I}} \|w_k - w\| \to 0 \quad \text{as } n \to \infty. \end{aligned}$$

Hence $x_k \rightrightarrows x$, i.e. the set $\{x_n | \ n \in \mathbb{N}\} \subseteq C(\mathcal{I}, \mathbb{R}^{n_x})$ is relatively compact. We will show that $\{y_n | \ n \in \mathbb{N}\} \subseteq C(\mathcal{I}, \mathbb{R}^{n_y})$ is also relatively compact. With (5.1b) we have $y_n(t) = \psi_g(x_n(t), t)$ and the x_n are relatively compact in $C(\mathcal{I}, \mathbb{R}^{n_x})$, i.e. there is a subsequence x_k with $x_k \rightrightarrows x$. Define $y(t) = \psi_g(x(t), t)$ which is continuous. Then $y_k \rightrightarrows y$ because ψ_g is Lipschitz continuous. Hence the y_n are relatively compact in $C(\mathcal{I}, \mathbb{R}^{n_y})$. By the Arzelà-Ascoli-Theorem A.19 $\{x_n\}$ and $\{y_n\}$ are both equicontinuous. We are now able to show that $\{u_n\} \subseteq C(\mathcal{I}, V)$ is equicontinuous on $\mathcal{I}$. Let $L_{\mathcal{R}} > 0$ be the Lipschitz constant of $\mathcal{R}$, fix $\bar{t} \in \mathcal{I}$ and let $\varepsilon > 0$. Since r is continuous and $\{x_n\}$, $\{y_n\}$ are equicontinuous there is $\delta > 0$ (independent of n) such that

$$\|x_n(\bar{t}) - x_n(t)\| < \frac{\mu_{\mathcal{B}}}{3 L_{\mathcal{R}}} \varepsilon,$$

$$\|y_n(\bar{t}) - y_n(t)\| < \frac{\mu_{\mathcal{B}}}{3 L_{\mathcal{R}}} \varepsilon,$$

$$\|r(\bar{t}) - r(t)\| < \frac{\mu_{\mathcal{B}}}{3} \varepsilon$$

for all $t \in \mathcal{I}$ with $|t - \bar{t}| < \delta$. Using the strong monotonicity of $\mathcal{B}$ gives for all $t \in \mathcal{I}$:

$$\begin{aligned} \mu_{\mathcal{B}} \|u_n(t) - u_n(\bar{t})\|_V^2 &\leq \langle \mathcal{B}(u_n(t)) - \mathcal{B}(u_n(\bar{t})), u_n(t) - u_n(\bar{t}) \rangle \\ &\overset{(5.3c)}{=} \langle \widetilde{\mathcal{R}}(z_n(\bar{t})) - \widetilde{\mathcal{R}}(z_n(t)) + r(\bar{t}) - r(t), u_n(t) - u_n(\bar{t}) \rangle \\ &\leq \left(L_{\mathcal{R}} \|z_n(t) - z_n(\bar{t})\| + \|r(t) - r(\bar{t})\|_{V^*} \right) \|u_n(t) - u_n(\bar{t})\|_V \end{aligned}$$

Then we have

$$\|u_n(t) - u_n(\bar{t})\|_V \leq \frac{L_{\mathcal{R}}}{\mu_{\mathcal{B}}} \left(\|x_n(t) - x_n(\bar{t})\| + \|y_n(t) - y_n(\bar{t})\| \right) + \frac{1}{\mu_{\mathcal{B}}} \|r(t) - r(\bar{t})\|_{V^*} < \varepsilon.$$

We deduce that $\{u_n\}$ is equicontinuous on $\mathcal{I}$.

Ad (4). Since $\{x_n\}$ is relatively compact in $C(\mathcal{I}, \mathbb{R}^{n_x})$ there is a subsequence n' and $x_* \in C(\mathcal{I}, \mathbb{R}^{n_x})$ such that

$$x_{n'} \rightrightarrows x_* \quad \text{as } n' \to \infty.$$

With $y_*(t) := \psi(x_*(t), t)$ we have as before that

$$y_{n'} \rightrightarrows y_* \quad \text{as } n' \to \infty$$

and hence

$$z_{n'} \rightrightarrows z_* \quad \text{as } n' \to \infty.$$

However, for u_n we have not proven relative compactness. Our goal is to show pointwise convergence of $u_{n'}$ to a limit u_* and apply Theorem A.18. The spaces V and V^* are reflexive and $u_{n'}(t)$ and $\mathcal{B}(u_{n'}(t))$ are uniformly bounded due to the a priori estimates above. Then for every $t \in \mathcal{I}$ there is a subsequence of n', which we denote by n''_t to underline its dependence on t, and $\bar{u}_t \in V$, $\bar{w}_t \in V^*$ such that

$$u_{n''_t}(t) \rightharpoonup \bar{u}_t \text{ in } V, \quad \mathcal{B}(u_{n''_t}(t)) \rightharpoonup \bar{w}_t \text{ in } V^* \quad \text{as } n''_t \to \infty. \tag{5.7}$$

This is due to the Theorem of Eberlein and Šmuljan, cf. [Zei90a, Theorem 21.D]. We will show now for all $t \in \mathcal{I}$:

(i) $\langle \mathcal{R}(z_{n''_t}(t), t), u_{n''_t}(t) \rangle_V \to \langle \mathcal{R}(z_*(t), t), \bar{u}_t(t) \rangle_V$ as $n''_t \to \infty$.

(ii) $\langle \bar{w}_t, v \rangle_V + \langle \mathcal{R}(z_*(t), t), v \rangle_V = 0$ for all $v \in V$.

(iii) $\mathcal{B}(\bar{u}_t) = \bar{w}_t$.

Ad (i). It is

$$
\begin{aligned}
\langle \mathcal{R}(z_{n''_t}(t), t), u_{n''_t}(t) \rangle_V &= \langle \mathcal{R}(z_{n''_t}(t), t) - \mathcal{R}(z_*(t), t), u_{n''_t}(t) \rangle_V + \langle \mathcal{R}(z_*(t), t), u_{n''_t}(t) \rangle_V \\
&\to \langle \mathcal{R}(z_*(t), t), \bar{u}_t \rangle_V \quad \text{as } n''_t \to 0
\end{aligned}
$$

because $\mathcal{R}$ is Lipschitz continuous, $u_{n''_t}(t)$ is uniformly bounded and $z_{n''_t} \rightrightarrows z_*$. So the first term vanishes as $n''_t \to 0$ and the second term tends to $\langle \mathcal{R}(z_*(t), t), \bar{u}_t \rangle_V$ because $u_{n''_t}(t) \rightharpoonup \bar{u}_t$.

Ad (ii). Let $n''_t \geq k$, $k \in \mathbb{N}$ and $v \in V_k$. The map $w \mapsto \langle w, v \rangle_V$ is in V^{**} and so we have with (5.7) that

$$\langle \mathcal{B}(u_{n''_t}(t)), v \rangle_V \to \langle \bar{w}_t, v \rangle_V \quad \text{as } n''_t \to \infty$$

and

$$\langle \mathcal{R}(z_{n''_t}(t), t), v \rangle_V \to \langle \mathcal{R}(z_*(t), t), v \rangle_V \quad \text{as } n''_t \to \infty$$

because $z_{n''_t} \rightrightarrows z_*$ and $\mathcal{R}$ is Lipschitz continuous. From the Galerkin equations (5.3c) we know that

$$\langle \mathcal{B}(u_{n''_t}(t)), v \rangle_V + \langle \mathcal{R}(z_{n''_t}(t), t), v \rangle_V = 0$$

and letting $n''_t \to \infty$ we deduce

$$\langle \bar{w}_t, v \rangle_V + \langle \mathcal{R}(z_*(t), t), v \rangle_V = 0 \quad \forall v \in \bigcup_{k \in \mathbb{N}} V_k$$

Since $\bigcup_{k \in \mathbb{N}} V_k \subseteq V$ dense and $\mathcal{B}(u_{n'}(t))$, $\mathcal{R}(z_{n'}(t), t)$ are bounded this also holds for $v \in V$, cf. [Zei90a, Proposition 21.26 (c)].

Ad (iii). We have (5.3c) and additionally we only have to show

$$\lim_{n''_t \to \infty} \langle \mathcal{B}(u_{n''_t}(t)), u_{n''_t}(t) \rangle_V \leq \langle \bar{w}_t, \bar{u}_t \rangle_V$$

to apply the monotonicity trick (cf. [Zei90b, p.474 or p.778]) since $\mathcal{B}$ is hemicontinuous and monotone. It is

$$\langle \mathcal{B}(u_{n''_t}(t)), u_{n''_t}(t) \rangle_V \overset{(5.3c)}{=} -\langle \mathcal{R}(z_{n''_t}(t), t), u_{n''_t}(t) \rangle_V$$
$$\rightarrow -\langle \mathcal{R}(z_*(t), t), \bar{u}_t(t) \rangle_V \quad \text{as } n''_t \to \infty$$

with (i). Then (ii) gives (iii) with the monotonicity trick.

Having shown that we know that for given z_* and fixed $t \in \mathcal{I}$ all weakly convergent subsequences of n' fulfill in the limit

$$\mathcal{B}(\bar{u}_t) = \mathcal{R}(z_*(t), t) \quad \text{in } V^*. \tag{5.8}$$

The solution of (5.8) is unique and hence we have that

$$u_{n'}(t) \rightharpoonup \bar{u}_t \text{ in } V, \quad \mathcal{B}(u_{n'}(t)) \rightharpoonup \mathcal{B}(\bar{u}_t) \text{ in } V^* \quad \text{as } n' \to \infty$$

with [Zei86, Proposition 10.13 (4)]. We set $u_*(t) := \bar{u}_t$ and observe that $u_* \in C(\mathcal{I}, V)$ because for $t_k \to t$ we have

$$\|u_*(t_k) - u_*(t)\|_V \leq \frac{1}{\mu_\mathcal{B}} \|\mathcal{B}(u_*(t_k)) - \mathcal{B}(u_*(t))\|_{V^*}$$
$$\overset{(5.8)}{=} \frac{1}{\mu_\mathcal{B}} \|\mathcal{R}(z_*(t_k), t_k) - \mathcal{R}(z_*(t), t)\|_{V^*} \to 0 \quad \text{as } k \to \infty$$

because $\mathcal{B}$ is strongly monotone and z_* and $\mathcal{R}$ are continuous. Additionally we have

$$\langle \mathcal{B}(u_*(t)), u_*(t) - u_{n'}(t) \rangle_V \to 0, \tag{5.9}$$
$$\langle \mathcal{B}(u_{n'}(t)), u_*(t) \rangle_V \to \langle \mathcal{B}(u_*(t)), u_*(t) \rangle_V \tag{5.10}$$

for $n' \to 0$ and (i) also holds for the sequence n'. Using the strong monotonicity of $\mathcal{B}$ again we see that

$$
\begin{aligned}
&\mu_{\mathcal{B}} \|u_*(t) - u_{n'}(t)\|_V^2 \\
\leq\ & \langle \mathcal{B}(u_*(t)) - \mathcal{B}(u_{n'}(t)), u_*(t) - u_{n'}(t) \rangle_V \\
\overset{(5.3c)}{=}\ & \underbrace{\langle \mathcal{B}(u_*(t)), u_*(t) - u_{n'}(t) \rangle_V}_{\overset{(5.9)}{\to} 0} - \underbrace{\langle \mathcal{B}(u_{n'}(t)), u_*(t) \rangle_V}_{\overset{(5.10)}{\to} \langle \mathcal{B}(u_*(t)), u_*(t) \rangle_V} - \underbrace{\langle \mathcal{R}(z_{n'}(t), t), u_{n'}(t) \rangle_V}_{\overset{(i)}{\to} \langle \mathcal{R}(z_*(t), t), u_*(t) \rangle_V} \\
\overset{(5.8)}{\to}\ & 0 \quad \text{as } n' \to \infty.
\end{aligned}
$$

Hence we have that $u_{n'}(t) \to u_*(t)$ pointwise in V as $n' \to \infty$. We conclude now with Theorem A.18 that $u_{n'} \rightrightarrows u_*$ because $\{u_{n'}\}$ is equicontinuous as shown above.

Ad (5). It remains to show that (z_*, u_*) fulfills the original equations (5.1). For u_* this is clear from (5.8) and y_* is given by $y_*(t) = \psi_g(x_*(t), t)$. So we need to show that (5.1a) is fulfilled. Therefore integrating the Galerkin equations (5.3a) gives

$$
m(x_{n'}(t), t) = m(x_{n'}(t_0), t_0) - \int_{t_0}^t f(z_{n'}(s), u_{n'}(s), s) \mathrm{d}s
$$

Using the uniform convergence of $(x_{n'})$, $(z_{n'})$ and $(u_{n'})$, the Lipschitz properties of f and that $x_{n'}(t_0) = x_0$ we see that (5.1a) is fulfilled. This also means that $x_* \in C^1(\mathcal{I}, \mathbb{R}^{n_x})$ and with Theorem 5.1 the solution (z_*, u_*) is unique. Now we apply the convergence principle form [Zei86, Proposition 10.13 (1)] which states the following: Given a Banach space W, a sequence (w_n) in W and $w \in W$ fixed. If every subsequence of (w_n) has, in turn, a subsequence which converges to w (in the norm $\|\cdot\|_W$) then the original sequence converges to w, i.e. $\|w_n - w\|_W \to 0$ as $n \to \infty$. Here the space W is $C(\mathcal{I}, \mathbb{R}^{n_z} \times V)$ and the arguments above can be applied to any subsequence $\bar{n}$ with a corresponding subsequence $\bar{n}'$. Hence we have the convergence of the whole sequence. $\quad\square$

5.2. Application of the elliptic prototype in circuit simulation

In this section we present a simple coupled system which fits into the framework of the elliptic prototype discussed in the last section of this chapter. For literature on the underlying physics motivating the following cf. [Sim56, Tis04, Bau12]. We start with the MNA equations having been described in chapter 3 and choose one resistor to be treated separately. We denote it by $\mathcal{R}$ and we describe its geometry by $\Omega \subseteq \mathbb{R}^d$, $d \leq 3$, which is open and bounded. Its resistivity is implicitly given by the well-known Ohm's law

$$
\mathcal{J} = \sigma \mathcal{E}.
$$

Letting $\mathcal{I} := [t_0, T]$ be a fixed time interval it gives a relation between the current density $\mathcal{J} : \Omega \times \mathcal{I} \to \mathbb{R}^d$ and the electric field $\mathcal{E} : \Omega \times \mathcal{I} \to \mathbb{R}^d$ via the conductivity tensor $\sigma \in \mathbb{R}^{d \times d}$. Furthermore consider the current continuity equation

$$\nabla \cdot \mathcal{J} + \frac{\partial \rho}{\partial t} = 0$$

where $\rho : \Omega \times \mathcal{I} \to \mathbb{R}$ is the charge distribution. Neglecting any drift current, i.e. $\frac{\partial \rho}{\partial t} = 0$, and using the potential formulation $-\nabla \varphi = \mathcal{E}$, see [HW05], we obtain the Laplace equation

$$-\nabla \cdot \sigma \nabla \varphi = 0,$$

for the electrostatic potential $\varphi : \Omega \times \mathcal{I} \to \mathbb{R}$. We add mixed boundary conditions. More precisely, let the boundary $\Gamma := \partial \Omega$ be sufficiently smooth. Let $\Gamma = \Gamma_D \cup \Gamma_N$ be split with $\Gamma_D \cap \Gamma_N = \emptyset$ and $\mathrm{vol}_{d-1}(\Gamma_D) > 0$. By $\mathrm{vol}_{d-1}(\Gamma_D)$ we denote the $(d-1)$-dimensional measure of the set Γ_D. Furthermore we split the Dirichlet boundary as $\Gamma_D = \Gamma_c \cup \Gamma_{\bar{c}}$ with $\Gamma_c \cap \Gamma_{\bar{c}} = \emptyset$. Γ_c and $\Gamma_{\bar{c}}$ are the contacts of the device. We then formulate the Laplace equation with mixed boundary conditions:

$$-\nabla \cdot \sigma \nabla \varphi = 0, \quad \text{on } \Omega \times \mathcal{I}, \tag{5.11a}$$

$$\varphi|_{\Gamma_c} = v_{\mathrm{app}}, \quad \varphi|_{\Gamma_{\bar{c}}} = 0, \tag{5.11b}$$

$$\nabla \varphi \cdot \nu|_{\Gamma_N} = 0 \tag{5.11c}$$

for some applied voltage $v_{\mathrm{app}} : \mathcal{I} \to \mathbb{R}$. $\nu \in \mathbb{R}^3$ denotes the outward normal vector and with $x \in \Omega$ and $t \in \mathcal{I}$ we denote the spatial and temporal dependency of φ respectively. We will omit these dependencies most of the time. Note also that in the formulation (5.11) the electrostatic potential φ is time dependent because the applied voltage v_{app} is. With the Gauss Theorem and equation (5.11a) it follows that

$$0 = \int_\Omega \nabla \cdot \mathcal{J} \mathrm{d}x = \int_\Gamma \mathcal{J} \cdot \nu \mathrm{d}\Gamma \overset{(5.11c)}{=} \underbrace{\int_{\Gamma_c} \mathcal{J} \cdot \nu \mathrm{d}\Gamma}_{=: \widetilde{j_\mathcal{R}}} + \underbrace{\int_{\Gamma_{\bar{c}}} \mathcal{J} \cdot \nu \mathrm{d}\Gamma}_{=: j_\mathcal{R}} = \widetilde{j_\mathcal{R}} + j_\mathcal{R}$$

where $j_\mathcal{R}$ and $\widetilde{j_\mathcal{R}}$ are the currents through the contacts. Thus we have charge conservation and so we can restrict ourself to one current

$$j_\mathcal{R} = j_\mathcal{R}(\varphi) = \int_{\Gamma_{\bar{c}}} -\sigma \nabla \varphi \cdot \nu \mathrm{d}\Gamma,$$

cf. e.g. [Tis04, chapter 3.2] in the case of semiconductor modeling. We introduce a corresponding incidence matrix $A_\mathcal{R} \in \mathbb{R}^{n_e \times 1}$ as

$$(A_\mathcal{R})_i := \begin{cases} 1, & \text{if the branch of } \mathcal{R} \text{ leaves node } i, \\ -1, & \text{if the branch of } \mathcal{R} \text{ enters node } i, \\ 0, & \text{else.} \end{cases}$$

The applied potential can be expressed as

$$v_{\mathrm{app}} = v_{\mathrm{app}}(t) = A_R^\top e(t)$$

with e being the vector of node potentials, cf. chapter 3.
Together with the MNA equations we arrive at the coupled system:

$$A_C \frac{\mathrm{d}}{\mathrm{d}t} q_C(A_C^\top e, t) + \begin{pmatrix} A_R & A_{\mathcal{R}} \end{pmatrix} \begin{pmatrix} g_R(A_R^\top e, t) \\ j_{\mathcal{R}}(\varphi) \end{pmatrix} + A_L j_L + A_V j_V + A_I i_s(t) = 0 \quad (5.12\text{a})$$

$$\frac{\mathrm{d}}{\mathrm{d}t} \phi_L(j_L, t) - A_L^\top e = 0 \quad (5.12\text{b})$$

$$A_V^\top e - v_s(t) = 0 \quad (5.12\text{c})$$

$$j_{\mathcal{R}}(\varphi) + \int_{\Gamma_{\bar{c}}} \sigma \nabla \varphi \cdot \nu \mathrm{d}\Gamma = 0 \quad (5.12\text{d})$$

$$-\nabla \cdot \sigma \nabla \varphi = 0 \quad (5.12\text{e})$$

$$\varphi|_{\Gamma_c} = A_{\mathcal{R}}^\top e, \ \varphi|_{\Gamma_{\bar{c}}} = 0 \quad (5.12\text{f})$$

$$\nabla \varphi \cdot \nu|_{\Gamma_N} = 0 \quad (5.12\text{g})$$

Here (5.12e) is given on $\Omega \times \mathcal{I}$. We now formulate the variational version of (5.12). Therefore we assume to have a function $h \in H^1(\Omega)$ with $h|_{\Gamma_c} = 1$ and $h|_{\Gamma_{\bar{c}}} = 0$. We set

$$u := \varphi - h v_{\mathrm{app}}$$

and let

$$v \in V := \left\{ v \in H^1(\Omega) \middle| \ v|_{\Gamma_D} = 0 \right\}. \quad (5.13)$$

which is a real separable Hilbert space. Then we obtain the variational formulation of (5.11):

$$\int_\Omega \sigma \nabla u \nabla v \mathrm{d}x = -v_{\mathrm{app}} \int_\Omega \sigma \nabla h \nabla v \mathrm{d}x \quad \forall v \in V$$

We define the operators $\mathcal{B} : V \to V^*$ and $\mathcal{R} : \mathbb{R} \to V^*$ as follows:

$$\langle \mathcal{B}(u), v \rangle_V := \int_\Omega \sigma \nabla u \nabla v \mathrm{d}x \quad \forall u, v \in V \quad \text{and} \quad (5.14\text{a})$$

$$\langle \mathcal{R}(z), v \rangle_V := z \int_\Omega \sigma \nabla h \nabla v \mathrm{d}x \quad \forall z \in \mathbb{R}, \ v \in V \quad (5.14\text{b})$$

The current j_R can be rewritten as

$$
\begin{aligned}
j_R = -\widetilde{j_R} &= \int_{\Gamma_c} \sigma \nabla \varphi \cdot \nu \mathrm{d}\Gamma \\
&= \int_\Gamma h\sigma \nabla \varphi \cdot \nu \mathrm{d}\Gamma \\
&= \int_\Omega \nabla \cdot (h\sigma \nabla \varphi) \mathrm{d}x \\
&= \int_\Omega \underbrace{\nabla \cdot (\sigma \nabla \varphi)}_{=0} h + \sigma \nabla \varphi \nabla h \mathrm{d}x
\end{aligned}
$$

and so j_R depends on u and v_{app}:

$$
j_R = j_R(u, v_{\mathrm{app}}) = \int_\Omega \sigma \nabla(u + h v_{\mathrm{app}}) \nabla h \mathrm{d}x \tag{5.15}
$$

Lemma 5.4.
Let V be defined as in (5.13). Let $\mathcal{B}$, $\mathcal{R}$ and j_R be given as in (5.14) and (5.15), respectively. Let $\sigma \in \mathbb{R}^{d \times d}$ be positive definite. Then

(i) $\mathcal{B}$ is linear, bounded and strongly monotone,

(ii) $\mathcal{R}$ is linear and continuous,

(iii) j_R is Lipschitz continuous w.r.t. u and v_{app}.

PROOF:
(i) The linearity is clear. Let $u, v \in H^1(\Omega)$ and with the Cauchy-Schwarz inequality it follows that

$$
\begin{aligned}
\langle \mathcal{B}(u), v \rangle_V &= \int_\Omega \sigma \nabla u \nabla v \mathrm{d}x \\
&\leq \|\sigma\|_* \left(\int_\Omega \sum_{i=1}^d |\partial_i u|^2 \mathrm{d}x \right)^{\frac{1}{2}} \left(\int_\Omega \sum_{i=1}^d |\partial_i v|^2 \mathrm{d}x \right)^{\frac{1}{2}} \\
&\leq \|\sigma\|_* \|u\|_{H^1(\Omega)} \|v\|_{H^1(\Omega)}.
\end{aligned}
$$

Furthermore we have for $u \in V$ that

$$
\int_\Omega \sigma \nabla u \nabla u \mathrm{d}x \geq \mu_\sigma \int_\Omega \|\nabla u\|^2 \mathrm{d}x = \mu_\sigma \int_\Omega \sum_{i=1}^d |\partial_i u|^2 \mathrm{d}x
$$

for some $\mu_\sigma > 0$ since σ is positive definite and therefore $(\sigma y)^\top y \geq \mu_\sigma \|y\|^2$ for all $y \in \mathbb{R}^d$. From the Poincaré-Friedrich inequality we have a $c > 0$ such that

$$\|u\|_{L_2(\Omega)}^2 \leq c \left(\int_\Omega \sum_{i=1}^d |\partial_i u|^2 \, \mathrm{d}x + \underbrace{\left| \int_{\Gamma_D} u \mathrm{d}\Gamma \right|^2}_{=0,\ \text{since}\ u \in V} \right),$$

since the boundary Γ is sufficiently smooth and $\mathrm{vol}_{d-1}(\Gamma_D) > 0$, cf. [GGZ74, p.36]. We conclude

$$\int_\Omega \sigma \nabla u \nabla u \mathrm{d}x \geq \mu_\mathcal{R} \|u\|_{H^1(\Omega)}^2$$

for some $\mu_\mathcal{R} > 0$.

(ii) The linearity is obvious and for $z \in \mathbb{R}$, $v \in V$ we have as before that

$$\langle \mathcal{R}(z), v \rangle_V \leq |z| \, \|\sigma\|_* \, \|h\|_{H^1(\Omega)} \, \|v\|_{H^1(\Omega)} \, .$$

So $\mathcal{R}$ is bounded and therefore continuous.

(iii) Let be u, $\bar{u} \in V$ and $v_{\mathrm{app}} \in \mathbb{R}$. Then we conclude as before:

$$|j_\mathcal{R}(u, v_{\mathrm{app}}) - j_\mathcal{R}(\bar{u}, v_{\mathrm{app}})| = \int_\Omega \sigma \nabla (u - \bar{u}) \nabla h \mathrm{d}x \leq \|\sigma\|_* \, \|u - \bar{u}\|_V \, \|h\|_{H^1(\Omega)}$$

The Lipschitz continuity of $j_\mathcal{R}$ w.r.t. v_{app} follows analogously. $\qquad \square$

We reformulate system (5.12):

$$A_C \frac{\mathrm{d}}{\mathrm{d}t} q_C(A_C^\top e, t) + \begin{pmatrix} A_R & A_\mathcal{R} \end{pmatrix} \begin{pmatrix} g_R(A_R^\top e, t) \\ j_\mathcal{R}(u, A_\mathcal{R}^\top e) \end{pmatrix} + A_L j_L + A_V j_V + A_I i_s(t) = 0 \quad (5.16a)$$

$$\frac{\mathrm{d}}{\mathrm{d}t} \phi_L(j_L, t) - A_L^\top e = 0 \quad (5.16b)$$

$$A_V^\top e - v_s(t) = 0 \quad (5.16c)$$

$$\mathcal{B}(u) + \mathcal{R}(A_\mathcal{R}^\top e) = 0 \quad (5.16d)$$

The system (5.16) is the weak formulation of system (5.12) and can be analyzed within the framework of the elliptic prototype. The operator equation (5.16d) is an equation in V^*. The main results of the last section, namely the solvability theorem (Theorem 5.1), the perturbation estimate (Theorem 5.2) and the convergence result for the solutions of the corresponding Galerkin equations (Theorem 5.3), can be applied to (5.16) under suitable assumptions of the electrical network. This will be discussed briefly in the next remark.

Remark 5.5 (Applicaton of the prototype).

We require Assumptions 3.21, 3.22 and 3.26 to hold which assured global solvability and a perturbation estimate for the MNA equations (3.14), cf. Theorems 3.30 and 3.32. Furthermore we assume that $A_{\mathcal{R}}^{\top} Q_C = 0$ holds. This means that there is a path of capacitors connecting the nodes of the device, compare [Bod07, Corollary 4.6]. We perform the decoupling for the MNA equations, cf. Lemma 3.29. Condition $A_{\mathcal{R}}^{\top} Q_C = 0$ ensures that there is no dependency of u in the algebraic part, see the decoupling in (3.19). With the same notation as in Lemma 3.29 we obtain:

$$\frac{\mathrm{d}}{\mathrm{d}t} m \left(\begin{pmatrix} e_C \\ j_L \end{pmatrix}, t \right) - f \left(\begin{pmatrix} e_C \\ j_L \end{pmatrix}, e_{\overline{C}\overline{V}}, t \right) + \begin{pmatrix} \mathrm{p}_C^{\top} A_{\mathcal{R}} j_{\mathcal{R}}(u, A_{\mathcal{R}}^{\top} \mathrm{p}_C e_C) \\ 0 \end{pmatrix} = 0 \qquad (5.17\text{a})$$

$$g \left(\begin{pmatrix} e_C \\ j_L \end{pmatrix}, e_{\overline{C}\overline{V}}, t \right) = 0 \qquad (5.17\text{b})$$

$$\begin{pmatrix} e_{\overline{C}V} \\ j_V \end{pmatrix} - h \left(\begin{pmatrix} e_C \\ j_L \end{pmatrix}, e_{\overline{C}\overline{V}}, t \right) = 0 \qquad (5.17\text{c})$$

$$\mathcal{B}(u) + \mathcal{R}(A_{\mathcal{R}}^{\top} \mathrm{p}_C e_C) = 0 \qquad (5.17\text{d})$$

Then Lemma 5.4 together with Theorem 3.30 ensures that the assumptions concerning the involved functions are fulfilled. Hence we get the unique solution

$$(e, j_L, j_V, u) \in C(\mathcal{I}, \mathbb{R}^{n_e + n_L + n_V} \times V)$$

to (5.16) by Theorem 5.1 for a given initial value $(\mathrm{p}_C^{\top} e(t_0), j_L(t_0))^{\top} = (\mathrm{p}_C^{\top} e_0, j_{L0})^{\top}$. We have $(\mathrm{p}_C^{\top} e, j_L) \in C^1(\mathcal{I}, \mathbb{R}^{k_C + n_L})$. Similarly we obtain the corresponding Galerkin equations for (5.16) and (5.17) and can apply Theorem 5.3 if a suitable basis in V is taken.

If perturbations $\delta_e \in C(\mathcal{I}, \mathbb{R}^{n_e})$, $\delta_L \in C(\mathcal{I}, \mathbb{R}^{n_L})$, $\delta_V \in C(\mathcal{I}, \mathbb{R}^{n_V})$ and $\delta_u \in C(\mathcal{I}, V^*)$ are added to the right hand side of (5.16) we get:

$$A_C \frac{\mathrm{d}}{\mathrm{d}t} q_C(A_C^{\top} e, t) + \begin{pmatrix} A_R & A_{\mathcal{R}} \end{pmatrix} \begin{pmatrix} g_R(A_R^{\top} e, t) \\ j_{\mathcal{R}}(u, A_{\mathcal{R}}^{\top} e) \end{pmatrix} + A_L j_L + A_V j_V + A_I i_s(t) = \delta_e(t)$$

$$\frac{\mathrm{d}}{\mathrm{d}t} \phi_L(j_L, t) - A_L^{\top} e = \delta_L(t)$$

$$A_V^{\top} e - v_s(t) = \delta_V(t)$$

$$\mathcal{B}(u) + \mathcal{R}(A_{\mathcal{R}}^{\top} e) = \delta_u(t)$$

Now the decoupling proceeds as above and we obtain a similar system to (5.17) by just replacing the functions f, g, h with the functions f^{δ}, g^{δ}, h^{δ} from Theorem 3.32. Note that no perturbations of the MNA equations enter the operator equation because $A_{\mathcal{R}}^{\top} e = A_{\mathcal{R}}^{\top} \mathrm{p}_C e_C$. So Theorem 5.2 (in combination with Theorem 3.32) can be applied. Let $(e^{\delta}, j_L^{\delta}, j_V^{\delta}, u^{\delta})$ be the unique solution for a given initial value

$$(\mathrm{p}_C^{\top} e^{\delta}(t_0), j_L^{\delta}(t_0))^{\top} = (\mathrm{p}_C^{\top} e_0^{\delta}, j_{L0}^{\delta})^{\top} =: x_0^{\delta}$$

and let (e, j_L, j_V, u) be the solution for $(\delta_e, \delta_L, \delta_V, \delta_u) = 0$ with initial value

$$(p_C^\top e(t_0), j_L(t_0))^\top = (p_C^\top e_0, j_{L0})^\top =: x_0.$$

If $\|x_0 - x_0^\delta\|$ is suffiently small then there is $c > 0$ such that

$$\left\|e - e^\delta\right\|_\infty + \left\|j_L - j_L^\delta\right\|_\infty + \left\|j_V - j_V^\delta\right\|_\infty + \max_{t \in \mathcal{I}} \left\|u(t) - u^\delta(t)\right\|$$

$$\leq c \left(\|x_0 - x_0^\delta\| + \|\delta_e\|_\infty + \|\delta_L\|_\infty + \|\delta_V\|_\infty + \max_{t \in \mathcal{I}} \|\delta_u(t)\|_{V_*} \right).$$

Thus system (5.16) has Perturbation Index 1.

Remark 5.6 (Structure of the coupled system).
In Remark 5.5 we have seen how the results of the elliptic prototype can be applied to (5.16). Moreover the system (5.16) provides more structure than used by the prototype. We observe in (5.16d) that if u_1 is the solution to

$$\langle \mathcal{B}(u), v \rangle_V + \langle \mathcal{R}(1), v \rangle_V = 0,$$

i.e. $v_{\text{app}} = 1$, the solution to (5.16d) for an applied potential $v_{\text{app}} = A_\mathcal{R}^\top e$ is given by

$$u = v_{\text{app}} u_1. \tag{5.18}$$

Here u_1 is not time-dependent and so the time-dependency comes from v_{app} only. The solution formula (5.18) is due to the linearity of $\mathcal{B}$ and $\mathcal{R}$. If we define $\varphi_1 := u_1 + h$ we have that

$$\int_\Omega \sigma \nabla \varphi_1 \nabla v \mathrm{d}x = 0 \quad \forall v \in V.$$

Since $u_1 \in V$ we obtain

$$\int_\Omega \sigma \nabla \varphi_1 \nabla u_1 \mathrm{d}x = 0$$

and so

$$j_\mathcal{R} = \int_\Omega \sigma \nabla \varphi \nabla h \mathrm{d}x = v_{\text{app}} \int_\Omega \sigma \nabla \varphi_1 \nabla h \mathrm{d}x \stackrel{\varphi_1 = u_1 + h}{=} v_{\text{app}} \int_\Omega \sigma \nabla \varphi_1 \nabla \varphi_1 \mathrm{d}x.$$

For a sufficiently smooth solution $\varphi_1 = u_1 + h$ the conductivity

$$\rho := \int_\Omega \sigma \nabla \varphi_1 \nabla \varphi_1 \mathrm{d}x$$

is positive because σ is positive definite and $\nabla \varphi_1 \neq 0$ due to the boundary conditions. This means that $j_\mathcal{R}(u, A_\mathcal{R}^\top e) = \rho A_\mathcal{R}^\top e =: g_\mathcal{R}(A_\mathcal{R}^\top e)$ is strongly monotone in $A_\mathcal{R}^\top e$. With the device effectively becoming a linear resistor the solvability Theorem 3.30 for the MNA equations can be applied. Here the term $g_\mathcal{R}$ is allowed to appear in the algebraic part, i.e. the condition $A_\mathcal{R}^\top Q_C = 0$ is not necessary. Note that these arguments are not possible for the general prototype system, because there $\mathcal{B}$ and $\mathcal{R}$ may be nonlinear. For the construction of (hemicontinuous) strongly monotone operators (with a nonlinear $\sigma = \sigma(\nabla u, x)$) we refer to [GGZ74, pp. 68-70].

5.3. A parabolic prototype

In this section we discuss a parabolic prototype of a coupled system. Let $\mathcal{I} := [t_0, T]$ be an interval and $V \subseteq H \subseteq V^*$ be an evolution triple. Consider the system

$$x'(t) + f(z(t), u(t), t) = 0, \quad t \in \mathcal{I}, \tag{5.19a}$$

$$g(z(t), t) = 0, \tag{5.19b}$$

$$u'(t) + \mathcal{B}u(t) + \mathcal{R}(u(t), z(t), t) = 0, \quad \text{in } V^*, \tag{5.19c}$$

$$x(t_0) = x_0, \quad u(t_0) = u_0 \tag{5.19d}$$

with functions $f : \mathbb{R}^{n_z} \times H \times \mathcal{I} \to \mathbb{R}^{n_x}$, $g : \mathbb{R}^{n_z} \times \mathcal{I} \to \mathbb{R}^{n_y}$ and operators $\mathcal{B} : V \to V^*$ and $\mathcal{R} : V \times \mathbb{R}^{n_z} \times \mathcal{I} \to V^*$. As for the elliptic prototype the unknowns are $x(t) \in \mathbb{R}^{n_x}$, $y(t) \in \mathbb{R}^{n_y}$ and $u(t) \in V$ for $t \in \mathcal{I}$. The same convention $z(t) = (x(t), y(t))^\top$ as for the elliptic case is used, cf. section 5.1, and initial values $x_0 \in \mathbb{R}^{n_x}$ and $u_0 \in H$ are given. Note that equations (5.19a), (5.19b) represent a semi-linear (finite dimensional) DAE. In contrast to the elliptic case, (5.19c) is an (infinite dimensional) evolution equation involving a generalized derivative where a solution u will be in the space $W_2^1 = W_2^1(\mathcal{I}; V, H)$. Background material can be found in Appendix A.3 and A.5. The coupling of these two systems is realized by letting f depend on u and $\mathcal{R}$ depend on z. We will investigate system (5.19) regarding solvability, perturbation estimates and Galerkin convergence under appropriate assumptions. The unique solvability is obtained via a Galerkin approach. First we assemble the following assumptions.

Assumption 5.7.
The following assumptions hold for system (5.19):

(i) Let $\mathcal{I} := [t_0, T]$ be an interval and $V \subseteq H \subseteq V^*$ be an evolution triple.

(ii) The initial values $x_0 \in \mathbb{R}^{n_x}$, $u_0 \in H$ are given.

(iii) $f \in C(\mathbb{R}^{n_z} \times H \times \mathcal{I}, \mathbb{R}^{n_x})$ is Lipschitz continuous w.r.t. z and u.

(iv) $g \in C(\mathbb{R}^{n_z} \times \mathcal{I}, \mathbb{R}^{n_y})$ is uniquely solvable w.r.t. $y \in \mathbb{R}^{n_y}$, i.e. there is a solution function $\psi_g \in C(\mathbb{R}^{n_x} \times \mathcal{I}, \mathbb{R}^{n_y})$ such that $y = \psi_g(x, t)$ whenever $g((x, y)^\top, t) = 0$ for all x, y, t. Furthermore ψ_g is Lipschitz continuous w.r.t. x.

(v) $\mathcal{B} : V \to V^*$ is linear, strongly monotone and bounded.

(vi) $\mathcal{R} \in C(V \times \mathbb{R}^{n_z} \times \mathcal{I}, V^*)$ is monotone w.r.t. u, i.e.

$$\langle \mathcal{R}(u, z, t) - \mathcal{R}(\bar{u}, z, t), u - \bar{u} \rangle_V \geq 0 \quad \forall u, \bar{u} \in V, z \in \mathbb{R}^{n_z}, t \in \mathcal{I}$$

and Lipschitz continuous w.r.t. z. Furthermore there are constants $c_{\mathcal{R},1}, c_{\mathcal{R},2} > 0$ such that

$$\|\mathcal{R}(u, 0, t)\|_{V^*} \leq c_{\mathcal{R},1} \|u\|_V + c_{\mathcal{R},2} \quad \forall u \in V.$$

(vii) Let $\dim V = \infty$ and $\{v_1, v_2, \ldots\}$ be a basis of V in the sense of Definition A.11. Set $V_n := \{v_1, \ldots, v_n\}$ and let there be a sequence $(u_{n0}) \subseteq V$ with $u_{n0} \in V_n$ and $u_{n0} \to u_0$ in H as $n \to \infty$.

We remark here that the assumptions on the finite dimensional part (5.19a), (5.19b) are very similar to the assumptions required in Theorem 3.15. Especially sufficient conditions for (iv) are discussed in Remark 3.8. For proving unique solvability and convergence of the Galerkin solutions we proceed as follows. First, we show the uniqueness (Lemma 5.8) of a possible solution to (5.19). Then we prove a priori estimates for the Galerkin solutions (Lemma 5.9) and prove the unique solvability of the Galerkin equations (Lemma 5.10) which are given as follows:

$$x_n'(t) + f(z_n(t), u_n(t), t) = 0, \quad t \in \mathcal{I} \tag{5.20a}$$

$$g(z_n(t), t) = 0, \tag{5.20b}$$

$$\langle u_n'(t), v_i \rangle_V + \langle \mathcal{B} u_n(t), v_i \rangle_V + \langle \mathcal{R}(u_n(t), z_n(t), t), v_i \rangle_V = 0, \quad \text{for } i = 1, \ldots, n, \tag{5.20c}$$

$$x_n(t_0) = x_0, \quad u_n(t_0) = u_{n0}. \tag{5.20d}$$

The operator equation (5.20c) is formulated on the finite dimensional subspace $V_n \subseteq V$. So $u_n(t)$ is in V_n which also influences the finite dimensional variable z through the coupling. Hence $z_n = (x_n, y_n)^\top$, too, depends on the Galerkin step n. Finally, we will be able to prove solvability and convergence of the Galerkin solutions (Theorem 5.11).

Lemma 5.8 (Uniqueness).
Let Assumption 5.7 be fulfilled. If $(z, u) \in C(\mathcal{I}, \mathbb{R}^{n_z} \times H)$ with $x \in C^1(\mathcal{I}, \mathbb{R}^{n_x})$ and $u \in W_2^1(\mathcal{I}; V, H)$ is a solution to (5.19) then (z, u) is unique.

PROOF:
Let (z, u), $(\bar{z}, \bar{u})$ be two solutions to (5.19). We define $\Delta u := u - \bar{u}$, $\Delta z := z - \bar{z}$, $\Delta x := x - \bar{x}$ and $\Delta y := y - \bar{y}$. It is $\Delta u(t_0) = 0$ and since $u, \bar{u} \in W_2^1(\mathcal{I}; V, H)$ we can apply the integration by parts formula (Proposition A.10) and obtain

$$\frac{1}{2} \|\Delta u(t)\|_H^2 = \frac{1}{2} \|\Delta u(t)\|_H^2 - \frac{1}{2} \|\Delta u(t_0)\|_H^2$$

$$= \int_{t_0}^t \langle \Delta u'(s), \Delta u(s) \rangle \mathrm{d}s$$

$$= -\int_{t_0}^t \langle \mathcal{B} \Delta u(s), \Delta u(s) \rangle_V + \langle \mathcal{R}(u(s), z(s), s) - \mathcal{R}(\bar{u}(s), \bar{z}(s), s), \Delta u(s) \rangle_V \mathrm{d}s$$

$$\leq -\mu \int_{t_0}^t \|\Delta u(s)\|_V^2 \, \mathrm{d}s - \int_{t_0}^t \langle \mathcal{R}(u(s), z(s), s) - \mathcal{R}(\bar{u}(s), z(s), s), \Delta u(s) \rangle_V \mathrm{d}s$$

$$- \int_{t_0}^t \langle \mathcal{R}(\bar{u}(s), z(s), s) - \mathcal{R}(\bar{u}(s), \bar{z}(s), s), \Delta u(s) \rangle_V \mathrm{d}s$$

In the last line we used the strong monotonicity of $\mathcal{B}$ with $\mu > 0$. Since $\mathcal{R}$ is monotone in u and Lipschitz continuous in z we have that

$$-\langle \mathcal{R}(u(s), z(s), s) - \mathcal{R}(\bar{u}(s), z(s), s), \Delta u(s)\rangle_V \leq 0$$

and

$$\langle \mathcal{R}(\bar{u}(s), z(s), s) - \mathcal{R}(\bar{u}(s), \bar{z}(s), s), \Delta u(s)\rangle_V \leq L_{\mathcal{R}} \|\Delta z(s)\| \|\Delta u(s)\|_V$$
$$\leq \frac{\mu}{2} \|\Delta u(s)\|_V^2 + \frac{2L_{\mathcal{R}}^2}{\mu} \|\Delta z(s)\|^2$$

using the classical inequality (4.8). Hence

$$\|\Delta u(t)\|_H^2 + \mu \int_{t_0}^t \|\Delta u(s)\|_V^2 \, \mathrm{d}s \leq \frac{4L_{\mathcal{R}}^2}{\mu} \int_{t_0}^t \|\Delta z(s)\|^2 \, \mathrm{d}s$$

From this it can be concluded that

$$\|\Delta u(t)\|_H^2 \leq c_u \int_{t_0}^t \|\Delta x(s)\|^2 + \|\Delta y(s)\|^2 \, \mathrm{d}s \tag{5.21}$$

holds for a constant $c_u > 0$. For the algebraic part (5.19b) we have

$$y(t) = \psi_g(x(t), t), \qquad \bar{y}(t) = \psi_g(\bar{x}(t), t)$$

and using the Lipschitz continuity of ψ_g we observe that

$$\|\Delta y(t)\|^2 \leq c_y \|\Delta x(t)\|^2 \tag{5.22}$$

for a constant $c_y > 0$. Using that

$$2\Delta x(t)^\top \Delta x'(t) = \frac{\mathrm{d}}{\mathrm{d}t} \|\Delta x(t)\|^2$$

we see multiplying (5.19a) from the left by $\Delta x(t)^\top$ that

$$\frac{1}{2}\frac{\mathrm{d}}{\mathrm{d}t} \|\Delta x(t)\|^2 = \Delta x(t)^\top \left(f(\bar{z}(t), \bar{u}(t), t) - f(z(t), u(t), t)\right).$$

Integration over $[t_0, t]$ gives

$$\frac{1}{2} \|\Delta x(t)\|^2 = \frac{1}{2} \|\Delta x(t)\|^2 - \frac{1}{2} \|\Delta x(t_0)\|^2$$
$$= \int_{t_0}^t \Delta x(s)^\top \left(f(\bar{z}(s), \bar{u}(s), s) - f(z(s), u(s), s)\right) \mathrm{d}s$$

because $\Delta x(t_0) = 0$. With the Lipschitz continuity of f and the classical inequality (4.8) we estimate

$$2\Delta x(s)^\top \left(f(\bar z(s), \bar u(s), s) - f(z(s), u(s), s)\right)$$
$$\leq \ 2L_f \|\Delta x(s)\| \left(\|\Delta z(s)\| + \|\Delta u(s)\|_H\right)$$
$$\leq \ c_x \left(\|\Delta x(s)\|^2 + \|\Delta y(s)\|^2 + \|\Delta u(s)\|_H^2\right)$$

for scalars c_x, $L_f > 0$. Hence we get

$$\|\Delta x(t)\|^2 \leq c_x \int_{t_0}^t \|\Delta x(s)\|^2 + \|\Delta y(s)\|^2 + \|\Delta u(s)\|_H^2 \,\mathrm{d}s \qquad (5.23)$$

Now inserting (5.22) into (5.21), (5.23) and adding them gives

$$\|\Delta x(t)\|^2 + \|\Delta u(t)\|_H^2 \leq c_{xu} \int_{t_0}^t \|\Delta x(s)\|^2 + \|\Delta u(s)\|_H^2 \,\mathrm{d}s$$

for a constant $c_{xu} > 0$. An application of the Gronwall Lemma reveals that

$$\|\Delta x(t)\|^2 + \|\Delta u(t)\|_H^2 = 0$$

and we deduce that $\Delta x = 0$, $\Delta u = 0$ and with (5.22) also $\Delta y = 0$. $\qquad \square$

Lemma 5.9 (A priori estimates).
Let Assumption 5.7 be fulfilled. If $(z_n, u_n) \in C(\mathcal{I}, \mathbb{R}^{n_z} \times V_n)$ with $x_n \in C^1(\mathcal{I}, \mathbb{R}^{n_x})$ and $u_n \in W_2^1(\mathcal{I}; V, H)$ is a solution to the Galerkin equations (5.20), then there is a $C > 0$ such that

$$\max_{t \in \mathcal{I}} \|z_n(t)\| \leq C,$$
$$\max_{t \in \mathcal{I}} \|u_n(t)\|_H \leq C,$$
$$\|u_n\|_{L_2(\mathcal{I}, V)} \leq C,$$
$$\|w_n\|_{L_2(\mathcal{I}, V^*)} \leq C$$

where $w_n \in L_2(\mathcal{I}, V^*)$ is defined by

$$\langle w_n(t), v \rangle_V := \langle \mathcal{B} u_n(t), v \rangle_V + \langle \mathcal{R}(u_n(t), z_n(t), t), v \rangle_V, \quad \forall v \in V, \ n \in \mathbb{N}, \ t \in \mathcal{I}.$$

PROOF:
Let (z_n, u_n) be a solution of the Galerkin equations (5.20). Since $u_n \in W_2^1$ we can apply the integration by parts formula (Proposition A.10) and get

$$\frac{1}{2}\|u_n(t)\|_H^2 - \frac{1}{2}\|u_{n0}\|_H^2 \ = \ \int_{t_0}^t \langle u_n'(s), u_n(s) \rangle_V \mathrm{d}s$$
$$\overset{(5.20c)}{=} \ -\int_{t_0}^t \langle \mathcal{B}u_n(s), u_n(s) \rangle_V + \langle \mathcal{R}(u_n(s), z_n(s), s), u_n(s) \rangle_V \mathrm{d}s$$
$$\leq \ -\mu \int_{t_0}^t \|u_n(s)\|_V^2 \,\mathrm{d}s - \int_{t_0}^t \langle \mathcal{R}(0, z_n(s), s), u_n(s) \rangle_V \mathrm{d}s$$

using the strong monotonicity of $\mathcal{B}$ and the monotonicity of $\mathcal{R}$, i.e.

$$
\begin{aligned}
&-\langle \mathcal{R}(u_n(s), z_n(s), s), u_n(s) \rangle_V \\
= \ &-\langle \mathcal{R}(u_n(s), z_n(s), s) - \mathcal{R}(0, z_n(s), s), u_n(s) \rangle_V - \langle \mathcal{R}(0, z_n(s), s), u_n(s) \rangle_V \\
\leq \ &-\langle \mathcal{R}(0, z_n(s), s), u_n(s) \rangle_V.
\end{aligned}
$$

Furthermore we estimate

$$
\begin{aligned}
\langle \mathcal{R}(0, z_n(s), s), u_n(s) \rangle_V &= \langle \mathcal{R}(0, z_n(s), s) - \mathcal{R}(0, 0, s), u_n(s) \rangle_V + \langle \mathcal{R}(0, 0, s), u_n(s) \rangle_V \\
&\leq \left(L_\mathcal{R} \, \|z_n(s)\| + \|\mathcal{R}(0, 0, s)\|_{V^*} \right) \|u_n(s)\|_V \\
&\leq \frac{\mu}{2} \|u_n(s)\|_V^2 + \frac{2}{\mu} \left(L_\mathcal{R} \, \|z_n(s)\| + \|\mathcal{R}(0, 0, s)\|_{V^*} \right)^2 \\
&\leq \frac{\mu}{2} \|u_n(s)\|_V^2 + \frac{4}{\mu} \left(L_\mathcal{R}^2 \, \|z_n(s)\|^2 + \|\mathcal{R}(0, 0, s)\|_{V^*}^2 \right)
\end{aligned}
$$

for a constant $L_\mathcal{R} > 0$. Here we used the Lipschitz continuity of $\mathcal{R}$ w.r.t. z and the classical inequality (4.8). Hence we derive

$$
\frac{1}{2} \|u_n(t)\|_H^2 + \frac{\mu}{2} \int_{t_0}^t \|u_n(s)\|_V^2 \, ds \leq \frac{1}{2} \|u_{n0}\|_H^2 + \frac{4}{\mu} \int_{t_0}^t L_\mathcal{R}^2 \, \|z_n(s)\|^2 + \|\mathcal{R}(0, 0, s)\|_{V^*}^2 \, ds.
$$

Since $u_{n0} \to u_0$ in H as $n \to \infty$ the term $\|u_{n0}\|_H$ is bounded. The operator $\mathcal{R}$ is continuous and $\mathcal{I}$ compact, so $\|\mathcal{R}(0, 0, s)\|_{V^*}$ is bounded as well. We get the estimates

$$
\|u_n(t)\|_H^2 \leq c_{1,u,H} + c_{2,u,H} \int_{t_0}^t \|x_n(s)\|^2 + \|y_n(s)\|^2 \, ds \tag{5.24}
$$

and

$$
\|u_n\|_{L_2(\mathcal{I}, V)}^2 \leq c_{1,u,L_2} + c_{2,u,L_2} \|z_n\|_{L_2(\mathcal{I}, \mathbb{R}^{n_z})}^2 \tag{5.25}
$$

with constants $c_{1,u,H}, c_{2,u,H}, c_{1,u,L_2}, c_{2,u,L_2} > 0$. For the algebraic part (5.20b) we get

$$
\|y_n(t)\| = \|\psi_g(x_n(t), t)\| \leq L_{\psi_g} \, \|x_n(t)\| + \|\psi_g(0, t)\|
$$

for a $L_{\psi_g} > 0$ because of the Lipschitz continuity of ψ_g. ψ_g is continuous and $\mathcal{I}$ is compact, so $\|\psi_g(0, t)\|$ is bounded and we conclude that

$$
\|y_n(t)\|^2 \leq c_{1,y} + c_{2,y} \, \|x_n(t)\|^2 \tag{5.26}
$$

for constants $c_{1,y}, c_{2,y} > 0$. Using integration by parts and (5.20a) we observe

$$
\begin{aligned}
\frac{1}{2} \|x_n(t)\|^2 - \frac{1}{2} \|x_0\|^2 &= \int_{t_0}^t x_n(s)^\top x_n'(s) ds \\
&\leq \int_{t_0}^t \|x_n(s)\| \, \|f(z_n(s), u_n(s), s)\| \, ds \\
&\leq \int_{t_0}^t \bar{c}_1 \|x_n(s)\| \left(\|z_n(s)\| + \|u_n(s)\|_H + \|f(0, 0, s)\| \right) ds \\
&\leq \bar{c} \int_{t_0}^t \|x_n(s)\|^2 + \|z_n(s)\|^2 + \|u_n(s)\|_H^2 + \|f(0, 0, s)\|^2 \, ds
\end{aligned}
$$

for $\bar{c}_1, \bar{c} > 0$. Thus

$$\|x_n(t)\|^2 \le c_{1,x} + c_{2,x} \int_{t_0}^t \left(\|x_n(s)\|^2 + \|y_n(s)\|^2 + \left\| u_n(s)^2 \right\|_H \right) ds \qquad (5.27)$$

with constants $c_{1,x}$, $c_{2,x} > 0$ since f is continuous. Inserting (5.26) in (5.24), (5.27) and adding them gives

$$\|x_n(t)\|^2 + \|u_n(t)\|_H^2 \le c_{1,xu} + c_{2,xu} \int_{t_0}^t \|x_n(s)\|^2 + \|u_n(s)\|_H^2 \, ds$$

with constants $c_{1,xu}$, $c_{2,xu} > 0$. An application of the Gronwall Lemma reveals that there is a $\bar{C} > 0$ (independent of t and n) such that

$$\|x_n(t)\|^2 \le \bar{C}, \qquad \|u_n(t)\|_H^2 \le \bar{C}.$$

Applying this to (5.26) gives the desired bound on $\|y_n(t)\|$ and so $\|z_n(t)\|$ is bounded. Since z_n is uniformly bounded we also get the boundedness of $\|u_n\|_{L_2(\mathcal{I},V)}$ with (5.25). It still needs to be shown that $w_n \in L_2(\mathcal{I}, V^*)$ is bounded uniformly. We see

$$\begin{aligned}
\langle w_n(t), v \rangle_V &= \langle \mathcal{B}u_n(t), v \rangle_V + \langle \mathcal{R}(u_n(t), z_n(t), t), v \rangle_V \\
&\le (c_{\mathcal{R}} \|u_n(t)\|_V + L_{\mathcal{R}} \|z_n(t)\| + \|\mathcal{R}(u_n(t), 0, t)\|) \|v\|_V \\
&\le (c_{w,1} (\|u_n(t)\|_V + \|z_n(t)\|) + c_{w,2}) \|v\|_V
\end{aligned}$$

with constants $c_{w,1}$, $c_{w,2} > 0$ because $\|\mathcal{R}(u_n(t), 0, t)\| \le c_{\mathcal{R},1} \|u_n(t)\| + c_{\mathcal{R},2}$. Hence

$$\begin{aligned}
\|w_n\|_{L_2(\mathcal{I},V^*)} &= \int_{\mathcal{I}} \|w_n(t)\|_{V^*}^2 \, dt \\
&\le c_{w,L_2,1} \int_{\mathcal{I}} \|u_n(t)\|_V^2 + \|z_n(t)\|^2 \, dt + c_{w,L_2,2}
\end{aligned}$$

with $c_{w,L_2,1}$, $c_{w,L_2,2} > 0$. Since $\|u_n\|_{L_2(\mathcal{I},V)}$ and $\max_{t \in \mathcal{I}} \|z_n(t)\|$ are bounded we obtain the desired result. $\qquad\square$

Lemma 5.10 (Unique solvability of the Galerkin equations).
Let Assumption 5.7 be fulfilled. Then the Galerkin equations (5.20) have a unique solution $(z_n, u_n) \in C(\mathcal{I}, \mathbb{R}^{n_z} \times V_n)$ with $x_n \in C^1(\mathcal{I}, \mathbb{R}^{n_x})$ and $u_n \in W_2^1(\mathcal{I}; V, H)$.

PROOF:
Inserting the algebraic constraint (5.20b), reformulated as

$$y_n(t) = \psi_g(x_n(t), t),$$

into (5.20a) and (5.20c) we set

$$\widetilde{f}(x_n(t), u_n(t), t) := f((x_n(t), \psi_g(x_n(t), t))^\top, u_n(t), t), \qquad (5.28)$$

$$\widetilde{\mathcal{R}}(u_n(t), x_n(t), t) := \mathcal{R}(u_n(t), (x_n(t), \psi_g(x_n(t), t))^\top, t). \qquad (5.29)$$

We have $\widetilde{f} \in C(\mathbb{R}^{n_x} \times V_n \times \mathcal{I}, \mathbb{R}^{n_x})$ and $\widetilde{\mathcal{R}} \in C(V_n \times \mathbb{R}^{n_x} \times \mathcal{I}, V^*)$ because f, $\mathcal{R}$ and ψ_g are continuous. Considering now (5.20c) we proceed as for the Galerkin equations in chapter 4 and represent

$$u_n(t) = \sum_{j=1}^{n} \alpha_{nj}(t)v_j, \qquad u_{n0} = \sum_{j=1}^{n} \alpha_{nj}^0 v_j$$

with coefficients $\alpha_{nj}(t)$, $\alpha_{nj}^0 \in \mathbb{R}$, $t \in \mathcal{I}$. With (5.29) we have for $i = 1\ldots,n$:

$$\sum_{j=1}^{n} \alpha_{nj}'(t)(v_j|v_i)_H + \sum_{j=1}^{n} \alpha_{nj}(t)\langle \mathcal{B}v_j, v_i\rangle_V + \langle \widetilde{\mathcal{R}}(\sum_{j=1}^{n} \alpha_{nj}(t)v_j, x_n(t), t), v_i\rangle_V = 0$$

$$\sum_{j=1}^{n} \alpha_{nj}(t_0)v_j = \sum_{j=1}^{n} \alpha_{nj}^0 v_j$$

Setting

$$\alpha_n(t) := \begin{pmatrix} \alpha_{n1}(t) & \cdots & \alpha_{nn}(t) \end{pmatrix}^\top, \qquad \alpha_n^0 := \begin{pmatrix} \alpha_{n1}^0(t) & \cdots & \alpha_{nn}^0(t) \end{pmatrix}^\top$$

we write in matrix notation

$$G\alpha_n'(t) + B\alpha_n(t) + r(\alpha_n(t), x_n(t), t) = 0$$

with

$$G := ((v_j|v_i)_H)_{i,j=1,\ldots,n},$$
$$B := (\langle \mathcal{B}v_j, v_i\rangle_V)_{i,j=1,\ldots,n},$$
$$r(\alpha_n(t), x_n(t), t) := (\langle \widetilde{\mathcal{R}}(\sum_{j=1}^{n} \alpha_{nj}(t)v_j, x_n(t), t), v_i\rangle_V)_{i=1,\ldots,n}.$$

We have that $r \in C(\mathbb{R}^{n+n_x} \times \mathcal{I}, \mathbb{R}^n)$ because $\widetilde{\mathcal{R}}$ is continuous. Consider now the initial value problem

$$x_n'(t) = -\widetilde{f}(x_n(t), \sum_{j=1}^{n} \alpha_{nj}(t)v_j, t), \qquad x_n(t_0) = x_0 \qquad (5.30a)$$

$$\alpha_n'(t) = -G^{-1}(B\alpha_n(t) + r(\alpha_n(t), x_n(t), t)), \qquad \alpha_n(t_0) = \alpha_n^0 \qquad (5.30b)$$

which can be solved with the Peano Theorem in a neighborhood $J := [t_0, T_J) \subseteq \mathcal{I}$ of t_0. Let $(x_n^*, \alpha_n^*) \in C^1(J, \mathbb{R}^{n_x} \times \mathbb{R}^n)$ be this solution to (5.30). Then

$$u_n^*(t) := \sum_{j=1}^{n} \alpha_{jn}^*(t)v_j,$$
$$y_n^*(t) := \psi_g(x_n^*(t), t),$$
$$z_n^*(t) := \begin{pmatrix} x_n^*(t) & y_n^*(t) \end{pmatrix}^\top$$

solves (5.20). We have $u_n^* \in W_2^1$ and the initial value condition is fulfilled because the v_j, $j = 1, \ldots, n$, are linearly independent. Due to Lemma 5.9 we have that $\|x_n^*(t)\|$ and $\|u_n^*(t)\|_V$ are uniformly bounded by a constant $C > 0$. We set

$$\|\alpha_n^*(t)\|_{H_n} := \left\| \sum_{j=1}^n \alpha_{nj}^*(t) v_j \right\|_H = \|u_n^*(t)\|_H$$

and have

$$\|\alpha_n^*(t)\| \le C_{H_n} \|\alpha_n^*(t)\|_{H_n} \le C_{H_n} C$$

because the norm $\|\cdot\|_{H_n}$ is equivalent to $\|\cdot\|$ on $\mathbb{R}^n$ ($C_{H_n} > 0$). This is due to the fact that the v_j are linearly independent. So the solution (x_n^*, α_n^*) can be extended to the end of the interval, cf. [Zei90b, p.800]. The uniqueness follows using the same arguments as in Lemma 5.8. $\qquad\square$

Theorem 5.11 (Solvability and Galerkin convergence).
Let Assumption 5.7 be fulfilled. Then the original system (5.19) has a unique solution $(z, u) \in C(\mathcal{I}, \mathbb{R}^{n_z} \times H)$ with $x \in C^1(\mathcal{I}, \mathbb{R}^{n_x})$ and $u \in W_2^1(\mathcal{I}; V, H)$. Furthermore we have for the solution (z_n, u_n) of the Galerkin equations (5.20) that

$$\max_{t \in \mathcal{I}} \|z_n(t) - z(t)\| \to 0,$$
$$\max_{t \in \mathcal{I}} \|u_n(t) - u(t)\|_H \to 0,$$
$$\|u_n - u\|_{L_2(\mathcal{I},V)} \to 0$$

as $n \to \infty$.

PROOF:
The proof proceeds in several steps. We will first present the outline of the proof before proving the details. Therefore let (z_n, u_n) be the solution of the Galerkin equations (5.20) and w_n defined as in Lemma 5.9.
Step 1. We show that the sequence (x_n) is equicontinuous. Using then the uniform a priori estimates from Lemma 5.9 we apply the Arzelà-Ascoli Theorem A.19 and the Theorem of Eberlein and Šmuljan, cf. [Zei90a, Theorem 21.D]. So there exists a subsequence of $(x_{n'}, u_{n'})$ and $x \in C(\mathcal{I}, \mathbb{R}^{n_x})$, $u \in L_2(\mathcal{I}, V)$, $w \in L_2(\mathcal{I}, V^*)$ and $u_T \in H$ such that

$$x_{n'} \rightrightarrows x, \qquad\qquad u_{n'} \rightharpoonup u \text{ in } L_2(\mathcal{I}, V),$$
$$u_{n'}(T) \rightharpoonup u_T \text{ in } H, \qquad\qquad w_{n'} \rightharpoonup w \text{ in } L_2(\mathcal{I}, V^*)$$

as $n' \to \infty$. Furthermore we have $y_n(t) := \psi_g(x_n(t), t)$ and set $y(t) := \psi_g(x(t), t)$. We see due to the Lipschitz continuity of ψ_g that $y_{n'} \rightrightarrows y$ and so $z_{n'} \rightrightarrows z$.

Step 2. We show:
(2.I) The key equation

$$\phi(T)(u_T \mid v)_H - \phi(t_0)(u_0 \mid v)_H = -\int_{\mathcal{I}} \langle w(t), v \rangle_V \phi(t) \mathrm{d}t + \int_{\mathcal{I}} (u(t) \mid v)_H \phi'(t) \mathrm{d}t \qquad (5.31)$$

holds for all $v \in V$, $\phi \in C^\infty(\mathcal{I})$.
(2.II) The limits u, w and u_T satisfy

$$\langle u'(t), v \rangle_V + \langle w(t), v \rangle_V = 0 \quad \forall v \in V,$$
$$u(t_0) = u_0, \quad u(T) = u_T, \quad u \in W_2^1(\mathcal{I}; V, H).$$

(2.III) For the given limit $z \in C(\mathcal{I}, \mathbb{R}^{n_z})$ it is $w(t) = \mathcal{B}(u(t), z(t), t)$ for all $t \in \mathcal{I}$. So the limits u and z fulfill equation (5.19c).
Step 3. $u_{n'} \to u$ in $C(\mathcal{I}, H)$ as $n' \to \infty$.
Step 4. The limits (z, u) satisfy the complete system (5.19) and $x \in C^1(\mathcal{I}, \mathbb{R}^{n_x})$.
Step 5. The preceding argumentation was done for a subsequence n' of the original sequence n. The limits fulfill (5.19) and are unique because of Lemma 5.8. As in the proof of Theorem 5.3 we can apply the convergence principle from [Zei86, Proposition 10.13]. Hence we have the convergence of the whole sequence $((z_n, u_n))$ in $C(\mathcal{I}, \mathbb{R}^{n_z} \times H)$. Additionally we have the weak convergence of the complete sequence (u_n) in $L_2(\mathcal{I}, V)$.
Step 6. It holds $u_n \to u$ in $L_2(\mathcal{I}, V)$ as $n \to \infty$. This completes the outline of the proof.
Ad (1). Because of the a priori estimates from Lemma 5.9 and the Lipschitz continuity of f there is a $D > 0$ (independent of n) such that

$$\max_{t \in \mathcal{I}} \| f(z_n(t), u_n(t), t) \| \leq D.$$

Let $\varepsilon > 0$ and $\delta(\varepsilon) := \frac{\varepsilon}{D}$. Then for $t, \bar{t} \in \mathcal{I}$ with $|t - \bar{t}| < \delta(\varepsilon)$ we see integrating (5.20a) over $[t, \bar{t}]$ that

$$\| x_n(\bar{t}) - x_n(t) \| \leq \int_t^{\bar{t}} \| f(z_n(s), u_n(s), s) \| \, \mathrm{d}s \leq D \, |t - \bar{t}| < \varepsilon.$$

Ad (2.I). We now write n instead of n'. Let $\phi \in C^\infty(\mathcal{I})$, $v \in V_k$, $k \in \mathbb{N}$ fixed, $n \geq k$. Since u_n, $\phi v \in W_2^1$ the integration by parts formula (Proposition A.10) can be applied and we obtain

$$(u_n(T) \mid \phi(T)v)_H - (u_{n0} \mid \phi(t_0)v)_H \quad = \quad \int_{\mathcal{I}} \langle u_n'(t), \phi(t)v \rangle_V + \langle \phi'(t)v, u_n(t) \rangle_V \mathrm{d}t$$
$$\overset{(5.20c)}{=} \int_{\mathcal{I}} -\langle w_n(t), \phi(t)v \rangle_V + (u_n(t) \mid v)_H \phi'(t) \mathrm{d}t$$

Since $u_n(T) \rightharpoonup u_T$ and $u_{n0} \to u_0$ in H we have

$$(u_n(T) \mid v)_H \to (u_T \mid v)_H, \qquad (u_{n0} \mid v)_H \to (u_0 \mid v)_H$$

as $n \to \infty$. From $u_n \rightharpoonup u$ and $w_n \rightharpoonup w$ we deduce with the Hölder inequality that

$$\int_{\mathcal{I}} \langle w_n(t), \phi(t)v \rangle_V \mathrm{d}t \to \int_{\mathcal{I}} \langle w(t), \phi(t)v \rangle_V \mathrm{d}t$$

$$\int_{\mathcal{I}} (v \,|\, u_n(t))_H \phi'(t) \mathrm{d}t \to \int_{\mathcal{I}} (u(t) \,|\, v)_H \phi'(t) \mathrm{d}t$$

as $n \to \infty$ because the embedding $V \subseteq H$ is continuous. So equation (5.31) is fulfilled for all $v \in \bigcup_{k \in \mathbb{N}} V_k$ which is dense in V. With the usual density argument as in the proof of Theorem 4.24, (2.I), we verify (5.31) for all $v \in V$.

Ad (2.II). For $\phi \in C_0^\infty(\mathcal{I})$ and $v \in V$ we obtain

$$\int_{\mathcal{I}} \langle w(t), v \rangle_V \phi(t) \mathrm{d}t \overset{(5.31)}{=} \int_{\mathcal{I}} (u(t) \,|\, v)_H \phi'(t) \mathrm{d}t$$

and hence $u' \in L_2(\mathcal{I}, V^*)$ exists and $u' = -w$ with Proposition A.9. So $u \in W_2^1$ and integration by parts (Proposition A.10) with $\phi v \in W_2^1$ for $v \in V$ and $\phi \in C^\infty(\mathcal{I})$ reveals:

$$
\begin{aligned}
(u(T) \,|\, \phi(T)v)_H - (u(t_0) \,|\, \phi(t_0)v)_H &= \int_{\mathcal{I}} \langle u'(t), \phi(t)v \rangle_V + \langle \phi'(t)v, u(t) \rangle_V \mathrm{d}t \\
&= \int_{\mathcal{I}} -\langle w(t), \phi(t)v \rangle_V + (u(t) \,|\, v)_H \phi'(t) \mathrm{d}t \\
&\overset{(5.31)}{=} (u_T \,|\, \phi(T)v)_H - (u_0 \,|\, \phi(t_0)v)_H
\end{aligned}
$$

Appropriate choices of ϕ and a density argument as in Theorem 4.24, (2.II), reveal that $u(T) = u_T$ and $u(t_0) = u_0$.

Ad (2.III). We set $X := L_2(\mathcal{I}, V)$, then $X^* = L_2(\mathcal{I}, V^*)$. For the limit $z \in C(\mathcal{I}, \mathbb{R}^{n_z})$ we define

$$\widetilde{\mathcal{B}} : X \to X^*, \quad (\widetilde{\mathcal{B}}(\widetilde{u}))(t) := \mathcal{B}\widetilde{u}(t) + \mathcal{R}(\widetilde{u}(t), z(t), t), \quad \widetilde{u} \in X, \ t \in \mathcal{I}.$$

We also write $(\widetilde{\mathcal{B}}(\widetilde{u}))(t) = \widetilde{\mathcal{B}}(\widetilde{u}(t))$. As for the w_n in the proof of Lemma 5.9 it can be shown that $\widetilde{\mathcal{B}}(\widetilde{u}) \in X^*$ because z is bounded and $\widetilde{u} \in X$. We show:

(i) $\widetilde{\mathcal{B}}$ is strongly monotone,

(ii) $\widetilde{\mathcal{B}}$ is hemicontinuous,

(iii) $\widetilde{\mathcal{B}}(u_n) \rightharpoonup w$ as $n \to \infty$ and

(iv) $\widetilde{\mathcal{B}}(u) = w$.

Ad (i). Let $\widetilde{u}_1, \widetilde{u}_2 \in X$. Then

$$
\begin{aligned}
\langle \widetilde{\mathcal{B}}(\widetilde{u}_1) - \widetilde{\mathcal{B}}(\widetilde{u}_2), \widetilde{u}_1 - \widetilde{u}_2 \rangle_X &= \int_{\mathcal{I}} \langle \mathcal{R}(\widetilde{u}_1(t), z(t), t) - \mathcal{R}(\widetilde{u}_2(t), z(t), t), \widetilde{u}_1(t) - \widetilde{u}_2(t) \rangle_V \, \mathrm{d}t \\
&\quad + \int_{\mathcal{I}} \langle \mathcal{B}(\widetilde{u}_1(t) - \widetilde{u}_2(t)), \widetilde{u}_1(t) - \widetilde{u}_2(t) \rangle_V \, \mathrm{d}t \\
&\geq \mu_{\mathcal{B}} \|\widetilde{u}_1 - \widetilde{u}_2\|_X^2
\end{aligned}
$$

because $\mathcal{B}$ is strongly monotone and $\mathcal{R}$ is monotone w.r.t. u.

Ad (ii). We follow a standard argument here, cf. [Zei90b, chapter 30.3b. (IV)]. We first remark that

$$
\langle (\widetilde{\mathcal{B}}(\widetilde{u}))(t), v \rangle_V \leq \left(c_{\widetilde{\mathcal{B}},1} \left(\|\widetilde{u}(t)\|_V + \|z(t)\| \right) + c_{\widetilde{\mathcal{B}},2} \right) \|v\|_V
$$

for all $\widetilde{u} \in X$, $v \in V$ and $t \in \mathcal{I}$. This follows as in the proof of Lemma 5.9 for the w_n. Let now $\bar{u}, \bar{w}, \bar{v} \in X$, $t \in \mathcal{I}$ and $s_k \to s$ as $k \to \infty$ with $0 \leq s, s_k \leq 1$. We then have

$$
\left| \langle \widetilde{\mathcal{B}}(\bar{u}(t) + s_k \bar{v}(t)), \bar{w}(t) \rangle_V \right| \leq \left(c_{\widetilde{\mathcal{B}},1} \left(\|\bar{u}(t) + s_k \bar{v}(t)\|_V + \|z(t)\| \right) + c_{\widetilde{\mathcal{B}},2} \right) \|\bar{w}(t)\|_V \leq q(t)
$$

with

$$
q(t) := c_{q,1} \left(\|\bar{u}(t)\|_V + \|\bar{v}(t)\|_V + \|z(t)\| + c_{q,2} \right) \|\bar{w}(t)\|_V
$$

for constants $c_{q,1}, c_{q,2} > 0$ because $s_k \leq 1$. Therefore the majorant function q is integrable because $\bar{u}, \bar{v}, z$ and $\bar{w}$ are. Furthermore we have that

$$
\langle \widetilde{\mathcal{B}}(\bar{u}(t) + s_k \bar{v}(t)), \bar{w}(t) \rangle_V \to \langle \widetilde{\mathcal{B}}(\bar{u}(t) + s \bar{v}(t)), \bar{w}(t) \rangle_V \quad \text{as } k \to \infty
$$

because of the continuity of $\mathcal{B}$ and $\mathcal{R}$. From the principle of majorized convergence, cf. [Zei90b, p.1015], it follows that

$$
\begin{aligned}
\lim_{k \to \infty} \langle \widetilde{\mathcal{B}}(\bar{u} + s_k \bar{v}), \bar{w} \rangle_X &= \lim_{k \to \infty} \int_{\mathcal{I}} \langle \widetilde{\mathcal{B}}(\bar{u}(t) + s_k \bar{v}(t)), \bar{w}(t) \rangle_V \, \mathrm{d}t \\
&= \langle \widetilde{\mathcal{B}}(u + sv), w \rangle_X
\end{aligned}
$$

This shows the hemicontinuity of $\widetilde{\mathcal{B}}$.

Ad (iii). Let $h \in X^{**}$ and with the Hölder inequality (Proposition A.7) and the Lipschitz continuity of $\mathcal{R}$ ($L_{\mathcal{R}} > 0$) it follows that

$$
\begin{aligned}
\langle h, \widetilde{\mathcal{B}}(u_n) - w \rangle_{X^*} &\leq \|h\|_{X^{**}} \left\| \widetilde{\mathcal{B}}(u_n) - w_n \right\|_{X^*} + |\langle h, w_n - w \rangle_{X^*}| \\
&\leq \|h\|_{X^{**}} \left(\int_{\mathcal{I}} \|\mathcal{R}(u_n(t), z(t), t) - \mathcal{R}(u_n(t), z_n(t), t)\|_{V^*}^2 \, \mathrm{d}t \right)^{\frac{1}{2}} \\
&\quad + |\langle h, w_n - w \rangle_{X^*}| \\
&\leq \underbrace{\|h\|_{X^{**}} L_{\mathcal{R}} \left(\int_{\mathcal{I}} \|z(t) - z_n(t)\|^2 \, \mathrm{d}t \right)^{\frac{1}{2}}}_{\to 0 \text{ as } n \to \infty} + \underbrace{|\langle h, w_n - w \rangle_{X^*}|}_{\to 0 \text{ as } n \to \infty}
\end{aligned}
$$

because $z_n \rightrightarrows z$ and $w_n \rightharpoonup w$ in X^* as $n \to \infty$.

Ad (iv). We have $u_n \rightharpoonup u$ in X and $\widetilde{\mathcal{B}}(u_n) \rightharpoonup w$ in X^* as $n \to \infty$. Since $\widetilde{\mathcal{B}}$ is hemicontinuous and monotone it remains to show that

$$\varlimsup_{n \to \infty} \langle \widetilde{\mathcal{B}}(u_n), u_n \rangle_X \leq \langle w, u_n \rangle_X$$

and the fundamental monotonicity trick can be applied, cf. [Zei90b, p.474]. Then we can deduce that $\widetilde{\mathcal{B}}(u) = w$. Integration by parts and the Galerkin equations yield

$$\frac{1}{2} \|u_n(T)\|_H^2 - \frac{1}{2} \|u_n(t_0)\|_H^2$$

$$= \int_{\mathcal{I}} \langle u'_n(t), u_n(t) \rangle_V \mathrm{d}t$$

$$\overset{(5.20c)}{=} -\int_{\mathcal{I}} \langle w_n(t), u_n(t) \rangle_V \mathrm{d}t$$

$$= -\int_{\mathcal{I}} \langle (\widetilde{\mathcal{B}}(u_n))(t), u_n(t) \rangle_V + \langle \mathcal{R}(u_n(t), z_n(t), t) - \mathcal{R}(u(t), z(t), t), u_n(t) \rangle_V \mathrm{d}t.$$

We have that $u_n(t_0) \to u_0$ in H and $u_n(T) \rightharpoonup u(T)$ in H and hence

$$|u(T)| \leq \varliminf_{n \to \infty} |u_n(T)|,$$

cf. [Zei90a, Proposition 21.23 (c)]. Furthermore the Hölder inequality gives

$$\int_{\mathcal{I}} \langle \mathcal{R}(u_n(t), z_n(t), t) - \mathcal{R}(u_n(t), z(t), t), u_n(t) \rangle_V \mathrm{d}t$$

$$\leq L_{\mathcal{R}} \int_{\mathcal{I}} \|z_n(t) - z(t)\| \, \|u_n(t)\|_V \, \mathrm{d}t$$

$$\leq \bar{c} \|z_n - z\|_\infty \|u_n\|_X \to 0 \text{ as } n \to \infty$$

for $\bar{c} > 0$ because $\|u_n\|_X$ is bounded and $z_n \rightrightarrows z$. We conclude:

$$\varlimsup_{n \to \infty} \langle \widetilde{\mathcal{B}}(u_n), u_n \rangle_X \leq \frac{1}{2} \|u(t_0)\|_H^2 - \frac{1}{2} \|u(T)\|_H^2$$

$$= -\int_{\mathcal{I}} \langle u'(t), u(t) \rangle_V \mathrm{d}t$$

$$= \int_{\mathcal{I}} \langle w(t), u(t) \rangle_V \mathrm{d}t = \langle w, u \rangle_X$$

Ad (3). We now show the convergence of u_n to u in $C(\mathcal{I}, H)$. Remember that with n we still denote a subsequence of the original sequence. In analogy to the proof of Theorem 4.25 or the proof of [Zei90a, Theorem 23.A] there is a sequence (p_n) of polynomials $p_n : \mathcal{I} \to V_n$ with

$$p_n \to u \text{ in } W_2^1 \quad \text{as } n \to \infty \tag{5.32}$$

because $\bigcup_n V_n \subseteq V$ dense. The embedding $W_2^1 \subseteq C(\mathcal{I}, H)$ is continuous, so we have

$$\max_{t \in \mathcal{I}} \|u(t) - p_n(t)\|_H \le c_1 \|u - p_n\|_{W_2^1} \to 0 \quad \text{as } n \to \infty$$

with a $c_1 > 0$. So it suffices to show that

$$\max_{t \in \mathcal{I}} \|u_n(t) - p_n(t)\|_H \to 0 \quad \text{as } n \to \infty.$$

Clearly

$$\|u_n(t_0) - p_n(t_0)\|_H \le \|u_n(t_0) - u(t_0)\|_H + \|u(t_0) - p_n(t_0)\|_H$$
$$\le \|u_{n0} - u_0\|_H + \max_{t \in \mathcal{I}} \|u_n(t) - p_n(t)\|_H \to 0 \quad \text{as } n \to \infty$$

because $u_{n0} \to u_0$ in H. Similar as in Theorem 4.25 we show that

$$\max_{t \in \mathcal{I}} \|u_n(t) - p_n(t)\|_H^2 - \|u_n(t_0) - p_n(t_0)\|_H^2 \to 0 \quad \text{as } n \to \infty \qquad (5.33)$$

which then proves the convergence of u_n in $C(\mathcal{I}, H)$. It is

$$\langle u_n'(t) - u'(t), u_n(t) - p_n(t) \rangle_V$$
$$\overset{(5.20c)}{=} -\langle w_n(t) + u'(t), u_n(t) - p_n(t) \rangle_V$$
$$\overset{(5.19c)}{=} \langle w(t) - w_n(t), u_n(t) - p_n(t) \rangle_V$$
$$= \langle (\widetilde{\mathcal{B}}(u))(t) - (\widetilde{\mathcal{B}}(u_n))(t) + (\widetilde{\mathcal{B}}(u_n))(t) - w_n(t), u_n(t) - p_n(t) \rangle_V$$
$$= \underbrace{\langle (\widetilde{\mathcal{B}}(u))(t) - (\widetilde{\mathcal{B}}(u_n))(t), u_n(t) - u(t) \rangle_V}_{\le 0} + \underbrace{\langle (\widetilde{\mathcal{B}}(u))(t) - (\widetilde{\mathcal{B}}(u_n))(t), u(t) - p_n(t) \rangle_V}_{\le \left\| (\widetilde{\mathcal{B}}(u))(t) - (\widetilde{\mathcal{B}}(u_n))(t) \right\|_{V_*} \|u(t) - p_n(t)\|_V}$$
$$+ \underbrace{\langle (\widetilde{\mathcal{B}}(u_n))(t) - w_n(t), u_n(t) - p_n(t) \rangle_V}_{\le L_{\mathcal{R}} \|z(t) - z_n(t)\| \|u_n(t) - p_n(t)\|_V}$$
$$\le \left\| (\widetilde{\mathcal{B}}(u))(t) - (\widetilde{\mathcal{B}}(u_n))(t) \right\|_{V_*} \|u(t) - p_n(t)\|_V + L_{\mathcal{R}} \|z(t) - z_n(t)\| \|u_n(t) - p_n(t)\|_V$$

with $L_{\mathcal{R}} > 0$. Integration by parts gives

$$\frac{1}{2} \|u_n(t) - p_n(t)\|_H^2 - \frac{1}{2} \|u_n(t_0) - p_n(t_0)\|_H^2$$
$$= \int_{t_0}^t \langle u_n'(s) - p_n'(s), u_n(s) - p_n(s) \rangle \mathrm{d}s$$
$$= \int_{t_0}^t \langle u'(s) - p_n'(s), u_n(s) - p_n(s) \rangle + \langle u_n'(s) - u'(s), u_n(s) - p_n(s) \rangle \mathrm{d}s$$
$$\le \|u - p_n\|_{W_2^1} \|u_n - p_n\|_X + \left\| (\widetilde{\mathcal{B}}(u)) - (\widetilde{\mathcal{B}}(u_n)) \right\|_{X_*} \|u - p_n\|_{W_2^1}$$
$$+ L_{\mathcal{R}} \|z - z_n\|_\infty \|u_n - p_n\|_X$$

Since $u_n \rightharpoonup u$ and $p_n \to u$ in W_2^1 as $n \to \infty$ the sequences (u_n) and (p_n) are bounded in X. Furthermore $\widetilde{\mathcal{B}} : X \to X^*$ is bounded and hence the sequence $(\widetilde{\mathcal{B}}(u_n))$ is bounded with the same reasoning as for w_n. This implies that the terms

$$\|u_n - p_n\|_X , \quad \left\| (\widetilde{\mathcal{B}}(u)) - (\widetilde{\mathcal{B}}(u_n)) \right\|_{X^*}$$

are bounded. Finally, we see that the right hand side tends to zero because $z_n \rightrightarrows z$ and (5.32) holds.

Ad (4). We have already seen that equation (5.19c) is fulfilled by the limit u given the limit z. Furthermore we can rewrite (5.20a) as follows

$$x_n(t) = x_{n0} - \int_{t_0}^t f(z_n(s), u_n(s), s)\mathrm{d}s.$$

It is $x_{n0} = x_0$ and letting $n \to \infty$ we observe

$$\left\| x_{n0} - \int_{t_0}^t f(z_n(s), u_n(s), s)\mathrm{d}s - x_0 + \int_{t_0}^t f(z(s), u(s), s)\mathrm{d}s \right\|$$
$$\leq \; c_f \int_{t_0}^t \|z_n(s) - z(s)\| + \|u_n(s) - u(s)\|_H \, \mathrm{d}s$$
$$\leq \; c_f (t - t_0) \max_{s \in \mathcal{I}} \left(\|z_n(s) - z(s)\| + \|u_n(s) - u(s)\|_H \right) \to 0$$

with $c_f > 0$ using the Lipschitz continuity of f and (3). Since $x_n \rightrightarrows x$ the limits z, u satisfy

$$x(t) = x_0 + \int_{t_0}^t f(z(s), u(s), s)\mathrm{d}s \quad \forall t \in \mathcal{I}.$$

Hence (5.19a) is fulfilled and $x \in C^1(\mathcal{I}, \mathbb{R}^{n_x})$ because f is continuous. By the definition of y the algebraic equation (5.19b) is automatically satisfied.

Ad (6). With integration by parts (Proposition A.10) we get

$$\frac{1}{2} \|u(T) - u_n(T)\|_H^2 - \frac{1}{2} \|u(t_0) - u_n(t_0)\|_H^2 \quad = \quad \int_{\mathcal{I}} \langle [u(t) - u_n(t)]', u(t) - u_n(t) \rangle_V \mathrm{d}t$$
$$\overset{(5.19c)}{=} \quad -\langle \widetilde{\mathcal{B}}(u), u - u_n \rangle_X - \langle u_n', u - u_n \rangle_X$$

and

$$(u_n(T) | u(T))_H - (u_n(t_0) | u(t_0))_H \quad = \quad \int_{\mathcal{I}} \langle u_n'(t), u(t) \rangle_V + \langle u'(t), u_n(t) \rangle_V \mathrm{d}t$$
$$\overset{(5.19c)}{=} \quad \langle u_n', u \rangle_X - \langle \widetilde{\mathcal{B}}(u), u_n \rangle_X.$$

So we have

$$\frac{1}{2}\|u(T) - u_n(T)\|_H^2 = \frac{1}{2}\|u(t_0) - u_n(t_0)\|_H^2 - \langle \widetilde{\mathcal{B}}(u) + u_n', u - u_n \rangle_X \qquad (5.34)$$

and

$$\langle u_n', u \rangle_X = \langle \widetilde{\mathcal{B}}(u), u_n \rangle_X + (u_n(T)\,|\,u(T))_H - (u_n(t_0)\,|\,u(t_0))_H. \qquad (5.35)$$

For convenience we set

$$\bar{w}_n(t) := \mathcal{B}u(t) + \mathcal{R}(u(t), z_n(t), t), \quad \forall t \in \mathcal{I}.$$

With the strong monotonicity of $\mathcal{B}$ and the monotonicity of $\mathcal{R}$ we obtain

$$
\begin{aligned}
&\mu_{\mathcal{B}}\|u - u_n\|_X^2 \\
\leq\ & \int_{\mathcal{I}} \langle \mathcal{B}(u(t) - u_n(t)), u(t) - u_n(t) \rangle_V \mathrm{d}t \\
\leq\ & \int_{\mathcal{I}} \langle \mathcal{B}(u(t) - u_n(t)) + \mathcal{R}(u(t), z_n(t), t) - \mathcal{R}(u_n(t), z_n(t), t), u(t) - u_n(t) \rangle_V \mathrm{d}t \\
\leq\ & \langle \bar{w}_n - w_n, u - u_n \rangle_X + \frac{1}{2}\|u(T) - u_n(T)\|_H^2 \\
\overset{(5.34)}{=}\ & \langle \bar{w}_n - w_n - \widetilde{\mathcal{B}}(u) - u_n', u - u_n \rangle_X + \frac{1}{2}\|u(t_0) - u_n(t_0)\|_H^2 \\
\overset{(5.20c)}{=}\ & \langle \bar{w}_n - \widetilde{\mathcal{B}}(u), u - u_n \rangle_X - \langle w_n + u_n', u \rangle_X + \frac{1}{2}\|u(t_0) - u_n(t_0)\|_H^2 \\
\overset{(5.35)}{=}\ & \langle \bar{w}_n - \widetilde{\mathcal{B}}(u), u - u_n \rangle_X - \langle w_n, u \rangle_X - \langle \widetilde{\mathcal{B}}(u), u_n \rangle_X \\
& - (u_n(T)\,|\,u(T))_H + (u_n(t_0)\,|\,u(t_0))_H + \frac{1}{2}\|u(t_0) - u_n(t_0)\|_H^2
\end{aligned}
$$

With the Hölder inequality we see that

$$\langle \bar{w}_n - \widetilde{\mathcal{B}}(u), u - u_n \rangle_X \leq \bar{c}\,\|z_n - z\|_\infty\,\|u - u_n\|_X \to 0$$

with $\bar{c} > 0$ as $n \to \infty$ because $z_n \rightrightarrows z$ and (u_n) is bounded in X. Since $u_n \rightharpoonup u$ in X and $w_n \rightharpoonup w = \widetilde{\mathcal{B}}(u)$ in X^* we have

$$\langle w_n, u \rangle_X \to \langle \widetilde{\mathcal{B}}(u), u \rangle_X \quad \text{and} \quad \langle \widetilde{\mathcal{B}}(u), u_n \rangle_X \to \langle \widetilde{\mathcal{B}}(u), u \rangle_X \text{ as } n \to \infty.$$

Since $u_n(t_0) \to u(t_0)$ in H and $u_n(T) \rightharpoonup u(T)$ in H we have with the integration by parts formula (Proposition A.10) that

$$
\begin{aligned}
\lim_{n\to\infty} \mu_{\mathcal{B}}\|u - u_n\|_X^2 &= -2\langle \widetilde{\mathcal{B}}(u), u \rangle_X - \|u(T)\|_H^2 + \|u(t_0)\|_H^2 \\
&= -2\langle \widetilde{\mathcal{B}}(u), u \rangle_X - 2\langle u', u \rangle_X \overset{(5.19c)}{=} 0.
\end{aligned}
$$

So $u_n \to u$ in $L_2(\mathcal{I}, V)$ as $n \to \infty$. $\qquad \square$

Remark 5.12 (Algebraic part of (5.19)).
In the investigated system (5.19) the algebraic part is given by the function g which does not depend on u. Allowing g to depend on u in the form

$$g(z(t), u(t), t) = 0$$

instead of (5.19b) makes the problem more complex. Apart from having to solve the algebraic part w.r.t. y with a continuous solution function that is Lipschitz continuous in x and u, it is not obvious how to achieve uniform convergence of y_n for the Galerkin sequence. This was crucial in the proof of Theorem 5.11 to obtain the convergence of u_n in $C(\mathcal{I}, H)$. If, however, $\mathcal{R}(u, z, t) = \mathcal{R}(u, x, t)$ only depends on x instead of z completely, then the convergence of y_n is not important to get the convergence of u_n in $C(\mathcal{I}, H)$. We then conclude the uniform convergence by representing y_n with the solution function.

For obtaining a perturbation result for system (5.19) we study the perturbed system

$$x'(t) + f(z(t), u(t), t, \delta_x(t)) = 0, \quad t \in \mathcal{I}, \tag{5.36a}$$

$$g(z(t), t, \delta_y(t)) = 0, \tag{5.36b}$$

$$u'(t) + \mathcal{B}u(t) + \mathcal{R}(u(t), z(t), t) + \delta_u(t) = 0, \quad \text{in } V^*, \tag{5.36c}$$

$$x(t_0) = x_0^\delta, \quad u(t_0) = u_0^\delta \tag{5.36d}$$

for perturbations $\delta_x(t) \in \mathbb{R}^{n_x}$, $\delta_y(t) \in \mathbb{R}^{n_y}$ and $\delta_u(t) \in V^*$ and perturbed initial values $x_0^\delta \in \mathbb{R}^{n_x}$ and $u_0^\delta \in H$. We obtain the following result showing that the prototype system (5.19) has Perturbation Index 1.

Theorem 5.13 (Perturbation result).
Consider system (5.36) together with the following assumptions:

(i) *Let $\mathcal{I} := [t_0, T]$ be an interval and $V \subseteq H \subseteq V^*$ be an evolution triple.*

(ii) *The initial values $x_0, x_0^\delta \in \mathbb{R}^{n_x}$, $u_0, u_0^\delta \in H$ are given.*

(iii) *$f \in C(\mathbb{R}^{n_z} \times H \times \mathcal{I} \times \mathbb{R}^{n_x}, \mathbb{R}^{n_x})$ is Lipschitz continuous w.r.t. z, u and δ_x.*

(iv) *$g \in C(\mathbb{R}^{n_z} \times \mathcal{I} \times \mathbb{R}^{n_y}, \mathbb{R}^{n_y})$ is uniquely solvable w.r.t. $y \in \mathbb{R}^{n_y}$, i.e. there is a solution function $\psi_g \in C(\mathbb{R}^{n_x} \times \mathcal{I} \times \mathbb{R}^{n_y}, \mathbb{R}^{n_y})$ such that $y = \psi_g(x, t, \delta_y(t))$ whenever $g((x, y)^\top, t, \delta_y(t)) = 0$ for all x, y, t, δ_y. Furthermore ψ_g is Lipschitz continuous w.r.t. x and δ_y.*

(v) *$\mathcal{B} : V \to V^*$ is linear, strongly monotone and bounded.*

(vi) *$\mathcal{R} \in C(V \times \mathbb{R}^{n_z} \times \mathcal{I}, V^*)$ is monotone w.r.t. u, i.e.*

$$\langle \mathcal{R}(u, z, t) - \mathcal{R}(\bar{u}, z, t), u - \bar{u} \rangle_V \geq 0 \quad \forall u, \bar{u}, v \in V, z \in \mathbb{R}^{n_z}, t \in \mathcal{I}$$

and Lipschitz continuous w.r.t. z. Furthermore there are constants $c_{\mathcal{R},1}$, $c_{\mathcal{R},2} > 0$ such that

$$\|\mathcal{R}(u, 0, t)\|_{V^*} \leq c_{\mathcal{R},1} \|u\|_V + c_{\mathcal{R},2} \quad \forall u \in V.$$

(vii) Let $\dim V = \infty$ and $\{v_1, v_2, \dots\}$ be a basis of V in the sense of Definition A.11. Set $V_n := \operatorname{span}\{v_1, \dots, v_n\}$ and let there be two sequences $(u_{n0}), (u_{n0}^\delta) \subseteq V$ with $u_{n0}, u_{n0}^\delta \in V_n$ and

$$u_{n0} \to u_0 \text{ in } H, \quad u_{n0}^\delta \to u_0^\delta \text{ in } H \quad \text{as } n \to \infty.$$

(viii) $\delta_x \in C(\mathcal{I}, \mathbb{R}^{n_x})$, $\delta_y \in C(\mathcal{I}, \mathbb{R}^{n_y})$ and $\delta_u \in C(\mathcal{I}, V^*)$.

Then the perturbed system (5.36) has a unique solution $(z^\delta, u^\delta) \in C(\mathcal{I}, \mathbb{R}^{n_z} \times H)$ with $z^\delta = (x^\delta, y^\delta)^\top$, $x^\delta \in C^1(\mathcal{I}, \mathbb{R}^{n_x})$ and $u^\delta \in W_2^1(\mathcal{I}; V, H)$. Let (z, u) be the solution for $(\delta_x, \delta_y, \delta_u) = 0$ with initial values $x(t_0) = x_0$ and $u(t_0) = u_0$. Then there is a $C > 0$ such that

$$\left\|z - z^\delta\right\|_\infty + \max_{t \in \mathcal{I}} \left\|u(t) - u^\delta(t)\right\|_H + \left\|u - u^\delta\right\|_{L_2(\mathcal{I},V)}$$
$$\leq \; C \left(\left\|x_0 - x_0^\delta\right\| + \left\|u_0 - u_0^\delta\right\|_H + \left\|\delta_x\right\|_\infty + \left\|\delta_y\right\|_\infty + \max_{t \in \mathcal{I}} \left\|\delta_u(t)\right\|_{V^*} \right).$$

PROOF:
Solvability. For given perturbations δ_x, δ_y and δ_u we define

$$f^\delta(z(t), u(t), t) := f(z(t), u(t), t, \delta_x(t)),$$
$$g^\delta(z(t), t) := g(z(t), t, \delta_y(t)),$$
$$\mathcal{R}^\delta(u(t), z(t), t) := \mathcal{R}(u(t), z(t), t) + \delta_u(t)$$

for all $t \in \mathcal{I}$. Then the functions f^δ, g^δ and the operator $\mathcal{R}^\delta$ inherit all the properties from the functions f, g and the operator $\mathcal{R}$ and this makes Theorem 5.11 applicable. Hence we get the desired unique solution (z^δ, u^δ) for the initial values x_0^δ and $u_0^\delta \in H$. For $(\delta_x, \delta_y, \delta_u) = 0$ and initial values x_0 and $u_0 \in H$ we denote the solution by (z, u).
Perturbation estimate. Building the difference between the perturbed and the unperturbed operator equation gives for $t \in \mathcal{I}$:

$$(u^\delta - u)'(t) + \mathcal{B}(u^\delta(t) - u(t)) + \mathcal{R}(u^\delta(t), z^\delta(t), t) - \mathcal{R}(u(t), z(t), t) + \delta_u(t) = 0$$

135

Integration by parts yields

$$\frac{1}{2}\left\|u^\delta(t) - u(t)\right\|_H^2 - \frac{1}{2}\left\|u^\delta(t_0) - u(t_0)\right\|_H^2$$

$$= \int_{t_0}^t \langle u^\delta(s) - u(s), u^\delta(s) - u(s)\rangle_V \mathrm{d}s$$

$$= -\int_{t_0}^t \langle \mathcal{B}(u^\delta(s) - u(s)), u^\delta(s) - u(s)\rangle_V \mathrm{d}s$$

$$-\int_{t_0}^t \langle \mathcal{R}(u^\delta(s), z^\delta(s), s) - \mathcal{R}(u(s), z(s), s) + \delta_u(s), u^\delta(s) - u(s)\rangle_V \mathrm{d}s$$

With the strong monotonicity of $\mathcal{B}$ ($\mu_\mathcal{B} > 0$) and the Lipschitz continuity of $\mathcal{R}$ ($L_\mathcal{R} > 0$) we obtain

$$\frac{1}{2}\left\|u^\delta(t) - u(t)\right\|_H^2 + \mu_\mathcal{B}\left\|u^\delta - u\right\|_{L_2(\mathcal{I},V)}^2$$

$$\leq \frac{1}{2}\left\|u^\delta(t_0) - u(t_0)\right\|_H^2 + \underbrace{\int_{t_0}^t \langle \mathcal{R}(u(s), z(s), s) - \mathcal{R}(u^\delta(s), z(s), s), u^\delta(s) - u(s)\rangle_V \, \mathrm{d}s}_{\leq 0}$$

$$+ \int_{t_0}^t \langle \mathcal{R}(u^\delta(s), z(s), s) - \mathcal{R}(u^\delta(s), z^\delta(s), s) + \delta_u(s), u^\delta(s) - u(s)\rangle_V \mathrm{d}s$$

$$\leq \frac{1}{2}\left\|u^\delta(t_0) - u(t_0)\right\|_H^2 + \int_{t_0}^t \left(L_\mathcal{R}\left\|z(s) - z^\delta(s)\right\| + \|\delta_u(s)\|_{V_*}\right)\left\|u^\delta(s) - u(s)\right\|_V \mathrm{d}s$$

$$\leq \frac{1}{2}\left\|u^\delta(t_0) - u(t_0)\right\|_H^2 + \frac{\mu_\mathcal{B}}{2}\left\|u^\delta - u\right\|_{L_2(\mathcal{I},V)}^2$$

$$+ \frac{4}{\mu_\mathcal{B}}\int_{t_0}^t L_\mathcal{R}^2\left\|z(s) - z^\delta(s)\right\|^2 + \|\delta_u(s)\|_{V_*}^2 \, \mathrm{d}s.$$

In the last line we used the classical inequality (4.8). Hence there is a constant $c_1 > 0$ such that

$$\left\|u^\delta - u\right\|_{L_2(\mathcal{I},V)} \leq c_1 \max_{t \in \mathcal{I}}\left(\left\|z(t) - z^\delta(t)\right\| + \|\delta_u(t)\|_{V_*}\right). \tag{5.37}$$

and

$$\max_{t \in \mathcal{I}}\left\|u^\delta(t) - u(t)\right\|_H \leq c_1 \max_{t \in \mathcal{I}}\left(\left\|z(t) - z^\delta(t)\right\| + \|\delta_u(t)\|_{V_*}\right). \tag{5.38}$$

Furthermore we have from (5.36b) that

$$y^\delta(t) = \psi_g(x^\delta(t), t, \delta_y(t)) \quad \text{and} \quad y(t) = \psi_g(x(t), t, 0).$$

We obtain

$$\left\|y(t) - y^\delta(t)\right\| \leq \left\|\psi_g(x(t), t, 0) - \psi_g(x^\delta(t), t, 0)\right\| + \left\|\psi_g(x^\delta(t), t, 0) - \psi_g(x^\delta(t), t, \delta_y(t))\right\|$$

$$\leq c_y\left(\left\|x(t) - x^\delta(t)\right\| + \|\delta_y(t)\|\right)$$

for a constant $c_y > 0$. Integration of (5.36a) and the Lipschitz continuity of f gives

$$\left\| x^\delta(t) - x(t) \right\| \leq \left\| x_0^\delta - x_0 \right\| + c_x \int_{t_0}^t \left\| z^\delta(s) - z(s) \right\| + \left\| u^\delta(s) - u(s) \right\|_H + \left\| \delta_x(s) \right\| \, \mathrm{d}s$$

Inserting into each other, using (5.38) and the Gronwall Lemma gives the desired result.

□

5.4. Application of the parabolic prototype in circuit simulation

In this section we discuss a coupled system in circuit simulation where the prototype system (5.19) from the previous section can be applied. In the classical formulation of the MNA equations (3.14) heating effects of certain circuit elements are not included. Nevertheless it is well known that resistors, for example, may depend significantly on their temperature. Due to miniaturization in chip design heating effects become ever more important. Accordingly, the influence on the circuit's behavior has to be simulated as well. In [Bar04] a first coupled thermal-electric model was described which adds thermal effects to the circuit by means of an additional 1D heat equation. Furthermore, comprehensive information on various heating models for resistors and diodes is given. This approach has been extended to coupled systems involving semiconductors, cf. [BJ11], and 2D/3D heat diffusion effects, cf. [Cul09]. In the following we will present the system from [Cul09] and discuss how it fits into the framework of the parabolic prototype.

Description of the Model

We follow the description in [Cul09] and point out the main modeling aspects in order to understand the corresponding coupled system mathematically. For a deeper treatment, especially concerning the physical motivation or the numerical and technical realization, we refer to [CdF08, Cul09, ABCdF10]. Let $\Omega \subseteq \mathbb{R}^d$ ($d = 2, 3$) be an open and bounded set and $\mathcal{I} := [t_0, T]$ a time interval. We denote the temporal variable by $t \in \mathcal{I}$ and the space variable by $x \in \Omega$. The set Ω describes the physical region of the electrical circuit where thermal effects are simulated. The circuit elements are either modeled as thermally active or thermally inactive ones. Every thermally active element is associated with a subset $\Omega_k \subseteq \Omega$ with $k = 1, \ldots, K$ and K being the number of thermally active elements. We require

$$\mathring{\Omega}_k \neq \emptyset, \quad \overline{\Omega}_k \subseteq \Omega, \quad \overline{\Omega}_k \cap \overline{\Omega}_j = \emptyset \qquad k, j = 1, \ldots, K, \ k \neq j.$$

with Ω and Ω_k having sufficiently smooth boundaries Γ and Γ_k respectively. A junction temperature $\theta_k = \theta_k(t)$, $k = 1, \ldots, K$, is assigned to every thermally active element. Additionally, the Joule power densities for every region Ω_k are represented by $p_k = p_k(t)$,

$k = 1, \ldots, K$. The temperature field on $\Omega \times \mathcal{I}$ is denoted by $T(x,t)$ and T_{env} is the ambient temperature. We write in vector notation

$$\theta(t) := \big(\theta_1(t), \ldots, \theta_K(t)\big)^\top, \quad p(t) := \big(p_1(t), \ldots, p_K(t)\big)^\top, \quad t \in \mathcal{I}.$$

We set

$$\chi_k(x) := \begin{cases} 1, & x \in \Omega_k \\ 0, & \text{else} \end{cases}, \quad k = 1, \ldots, K$$

being the indicator function of the set Ω_k. Then the distributed temperature field $T(x,t)$ is linked to the junction temperatures by

$$\theta_k(t) = \frac{1}{|\Omega_k|}(T(\cdot,t)|\chi_k)_{L_2(\Omega)} = \frac{\int_{\Omega_k} T(x,t)\mathrm{d}x}{|\Omega_k|}, \quad k = 1, \ldots, K, \quad t \in \mathcal{I}. \tag{5.39}$$

This means that θ_k is the mean value of T over Ω_k. The Joule power dissipated by the thermally active elements is given by

$$p_k(t) = \frac{1}{|\Omega_k|}W_k(\theta_k(t), v_{\mathrm{app},k}(t)), \quad t \in \mathcal{I} \tag{5.40}$$

where the W_k are functions depending on the junction temperature θ_k and the applied voltage $v_{\mathrm{app},k}$. We elaborate a concrete expression for the W_k subsequently, cf. Assumption 5.17. Heat diffusion in the simulation domain Ω is described by the heat diffusion equation given by

$$\alpha \frac{\partial T(x,t)}{\partial t} + \mathcal{L}T(x,t) = \sum_{k=1}^{K} p_k(t)\chi_k(x) \quad \text{on } \Omega \times \mathcal{I} \tag{5.41}$$

with

$$\mathcal{L}T(x,t) := -\sum_{i,j=1}^{d} \partial_i(\kappa_{ij}\partial_j T(x,t)) + \hat{\alpha}T(x,t),$$

cf. Example 2.1. The factor $\alpha > 0$ accounts for the thermal capacitance of the material whereas the coefficients $\kappa_{ij} = \kappa_{ij}(x)$ account for possibly anisotropic heat diffusion. Also a reaction term with

$$\hat{\alpha} > 0, \quad \text{if } d = 2 \quad \text{and} \quad \hat{\alpha} = 0, \quad \text{if } d = 3$$

is embodied to model heat loss in the missing third dimension if $d = 2$. On the right hand side the power terms p_k are uniformly distributed over their corresponding physical region $\Omega_k \subseteq \Omega$. This corresponds to (5.39) where the junction temperatures θ_k are extracted as mean values. Further ways of distributing the power terms p_k and extracting the temperatures θ_k are possible using different distribution functions. For the one dimensional case more details can be found in [Bar04]. Furthermore the heat diffusion equation (5.41) is complemented by suitable boundary conditions. Here we assume to have Dirichlet boundary conditions, i.e.

$$T(x,t) = T_{env} \qquad \text{on } \Gamma \times \mathcal{I}. \tag{5.42}$$

In [Cul09] especially Robin boundary conditions are considered, but the case of Dirichlet boundary conditions is also treated. The heat equation (5.41) is linked to the MNA equations (3.14) by the junction temperatures $\theta(t)$. Therefore we first introduce a corresponding incidence matrix

$$A_T \in \mathbb{R}^{n_e \times K}, \quad A_T = \big(A_{T,1}, \ldots, A_{T,K} \big)$$

which is defined by

$$(A_T)_{ij} := \begin{cases} 1, & \text{if the branch of the thermally active element } j \text{ leaves node } i, \\ -1, & \text{if the branch of the thermally active element } j \text{ enters node } i, \\ 0, & \text{else.} \end{cases}$$

In addition we describe all the branch currents of the thermally active elements by

$$j_T(t) = g_T(A_T^\top e(t), t, \theta(t))$$

with a function $g_T : \mathbb{R}^K \times \mathcal{I} \times \mathbb{R}^K \to \mathbb{R}^K$. A concrete expression for g_T will be presented subsequently. Following the derivation of the MNA equations in chapter 3 we end up with the following system:

$$A_C \frac{\mathrm{d}}{\mathrm{d}t} q_C(A_C^\top e(t), t) + A_R g_R(A_R^\top e(t), t) + A_T g_T(A_T^\top e(t), t, \theta(t))$$
$$+ A_L j_L(t) + A_V j_V(t) + A_I i_s(t) = 0$$
$$\frac{\mathrm{d}}{\mathrm{d}t} \phi_L(j_L(t), t) - A_L^\top e(t) = 0$$
$$A_V^\top e(t) - v_s(t) = 0$$

We allow the junction temperatures to have an influence on resistive elements only. In [Cul09] or [Bar04] an influence on certain types of controlled current sources is allowed. The applied potential of the k-th thermally active element is given by

$$v_{\mathrm{app},k}(t) = A_{T,k}^\top e(t), \quad k = 1, \ldots, K. \tag{5.43}$$

Finally the fully coupled electro-thermal system can be formulated as follows, cf. [Cul09]:

$$A_C \frac{\mathrm{d}}{\mathrm{d}t} q_C(A_C^\top e(t), t) + A_R g_R(A_R^\top e(t), t) + A_T g_T(A_T^\top e(t), t, \theta(t))$$

$$+ A_L j_L(t) + A_V j_V(t) + A_I i_s(t) = 0$$

$$\frac{\mathrm{d}}{\mathrm{d}t} \phi_L(j_L(t), t) - A_L^\top e(t) = 0$$

$$A_V^\top e(t) - v_s(t) = 0$$

$$|\Omega_k| \theta_k(t) - (T(\cdot, t) | \chi_k)_{L_2(\Omega)} = 0 \qquad (5.44)$$

$$|\Omega_k| p_k(t) - W_k(\theta_k(t), A_{T,k}^\top e(t)) = 0$$

$$\alpha \frac{\partial T(x, t)}{\partial t} + \mathcal{L} T(x, t) - \sum_{k=1}^{K} p_k(t) \chi_k(x) = 0$$

$$T(x, t)|_{\Gamma \times \mathcal{I}} - T_{env} = 0$$

Note that the variables $e(t)$, $j_L(t)$, $j_V(t)$, $\theta(t)$ and $p(t)$ are only time dependent whereas $T(x, t)$ also depends on the spatial variable $x \in \Omega$.

Application of the prototype

We are now going to transform the system (5.44) into a variational formulation such that the results for the parabolic prototype system can be applied. First we set

$$V := H_0^1(\Omega), \quad H := L_2(\Omega)$$

and thus $V^* = H^{-1}(\Omega)$ and $V \subseteq H \subseteq V^*$ forms an evolution triple. We set

$$u(t)(x) = u(x, t) := T(x, t) - T_{env}.$$

With the standard procedure of homogenization, multiplying by a test function $v \in V$ and integrating over Ω we obtain the weak formulation of (5.41), (5.42):

$$\langle u'(t), v \rangle_V + \langle \mathcal{B} u(t), v \rangle_V + \langle \widetilde{\mathcal{W}}(p(t)), v \rangle_V = 0 \qquad \forall v \in V$$

with $\mathcal{B} : V \to V^*$ and $\widetilde{\mathcal{W}} : \mathbb{R}^K \to V^*$ defined by

$$\langle \mathcal{B} u(t), v \rangle_V := \frac{1}{\alpha} \int_\Omega \sum_{i,j=1}^{d} \kappa_{ij} \partial_j u(x, t) \partial_i v(x) + \hat{\alpha} u(x, t) v(x) \mathrm{d}x \qquad (5.45)$$

$$\langle \widetilde{\mathcal{W}}(p(t)), v \rangle_V := -\frac{1}{\alpha} \sum_{k=1}^{K} p_k(t) \int_\Omega \chi_k(x) v(x) \mathrm{d}x + \frac{1}{\alpha} \int_\Omega \hat{\alpha} T_{env} v(x) \mathrm{d}x \qquad (5.46)$$

for all $v \in V$ and $t \in \mathcal{I}$. The coupling condition (5.39) has to be expressed in terms of u as well and we arrive at

$$\theta_k(t) = \frac{1}{|\Omega_k|}(u(\cdot,t)|\chi_k)_H + T_{env} =: \mathcal{K}_k(u(t)), \quad k = 1, \dots, K. \tag{5.47}$$

where $\mathcal{K}_k : H \to \mathbb{R}$. In vector notation we write

$$\theta(t) = \mathcal{K}(u(t)) := (\mathcal{K}_k(u(t)))_{k=1,\dots,K} \tag{5.48}$$

where $\mathcal{K} : H \to \mathbb{R}^K$. We fix the following simple result.

Lemma 5.14 (Properties of $\mathcal{K}_k$).
Let $\mathcal{K}_k$, $\mathcal{K}$ be given by (5.47) for $1 \leq k \leq K$ and (5.48) respectively. Then $\mathcal{K}_k$ and $\mathcal{K}$ are Lipschitz continuous.

PROOF:
Let be $u, \bar{u} \in H$. We then see that

$$|\mathcal{K}_k(u) - \mathcal{K}_k(\bar{u})| = |\Omega_k|^{-1} |(u - \bar{u}|\chi_k)_H| \leq |\Omega_k|^{-1} \|\chi_k\|_H \|u - \bar{u}\|_H$$

due to the Cauchy-Schwarz inequality for $1 \leq k \leq K$. The Lipschitz continuity of $\mathcal{K}_k$ and $\mathcal{K}$ follows directly. $\qquad\square$

We can rewrite the second coupling condition (5.40) with (5.43) as

$$p_k(t) = \frac{1}{|\Omega_k|}W_k(\theta_k(t), A_{T,k}^\top e(t)) = \frac{1}{|\Omega_k|}W_k(\mathcal{K}_k(u(t)), A_{T,k}^\top e(t))$$

and inserting into (5.46) gives

$$\begin{aligned}
\langle \widetilde{\mathcal{W}}(p(t)), v \rangle_V &= -\frac{1}{\alpha}\int_\Omega \sum_{k=1}^K \frac{1}{|\Omega_k|}W_k(\mathcal{K}_k(u(t)), A_{T,k}^\top e(t))\chi_k(x)v(x)\mathrm{d}x + \frac{1}{\alpha}\int_\Omega \hat{a}T_{env}v(x)\mathrm{d}x \\
&= -\sum_{k=1}^K \frac{1}{\alpha|\Omega_k|}W_k(\mathcal{K}_k(u(t)), A_{T,k}^\top e(t))(\chi_k|v)_{L_2(\Omega)} + \frac{1}{\alpha}(\hat{a}T_{env}|v)_{L_2(\Omega)}.
\end{aligned}$$

Furthermore we set

$$\begin{aligned}
&\langle \mathcal{W}(u(t), e(t)), v \rangle_V \\
&:= -\sum_{k=1}^K \frac{1}{\alpha|\Omega_k|}W_k(\mathcal{K}_k(u(t)), A_{T,k}^\top e(t))(\chi_k|v)_{L_2(\Omega)} + \frac{1}{\alpha}(\hat{a}T_{env}|v)_{L_2(\Omega)}.
\end{aligned} \tag{5.49}$$

With the given notation the weak (and transformed) formulation of system (5.44) reads

$$A_C \frac{\mathrm{d}}{\mathrm{dt}} q_C(A_C^\top e(t), t) + A_R g_R(A_R^\top e(t), t) + A_T g_T(A_T^\top e(t), t, \mathcal{K}(u(t)))$$

$$+ A_L j_L(t) + A_V j_V(t) + A_I i_s(t) = 0$$

$$\frac{\mathrm{d}}{\mathrm{dt}} \phi_L(j_L(t), t) - A_L^\top e(t) = 0 \tag{5.50}$$

$$A_V^\top e(t) - v_s(t) = 0$$

$$u'(t) + \mathcal{B}u(t) + \mathcal{W}(u(t), e(t)) = 0$$

Note that the last equation is an operator equation in V^* and $u'(t)$ denotes a generalized derivative. System (5.50) has to be supplemented by appropriate initial conditions which we will discuss later. In the following we are going to apply the solvability result (Theorem 5.11) and the perturbation result (Theorem 5.13) to the system (5.50). We will also omit the explicit dependence on t of the variables e, j_L, j_V and u.

First we present concrete expressions for the function g_T and the operator $\mathcal{W}$. We start with the following assumption.

Assumption 5.15.
Let $R_{0k} > 0$, $\theta_{0k} \in \mathbb{R}$, $\alpha_k > 0$ be given for $k = 1, \ldots, K$ and let be $\beta_k > -\frac{1}{\alpha_k} + \theta_{0k}$. Then we define

$$\widetilde{G}_k(\theta_k) := \frac{1}{R_{0k}(1 + \alpha_k(\theta_k - \theta_{0k}))} \tag{5.51}$$

and

$$G_k(\theta_k) := \begin{cases} \widetilde{G}_k(\beta_k), & \text{if } \theta_k < \beta_k, \\ \widetilde{G}_k(\theta_k), & \text{else.} \end{cases} \tag{5.52}$$

Furthermore we write

$$g_T(A_T^\top e, t, \theta) = G(\theta) A_T^\top e, \quad G(\theta) = \operatorname{diag} \{G_1(\theta_1), \ldots, G_K(\theta_K)\}. \tag{5.53}$$

Equation (5.51) yields a common model for the temperature depending conductivity of the k-th element where $R_{0k} > 0$ is the resistivity at a reference temperature θ_{0k} and $\alpha_k > 0$ is the temperature coefficient, see [Bar04]. So for constant temperature linear resistive behavior is assumed whereas the temperature depending behavior is nonlinear. The modified term allows us to prove useful properties.

Lemma 5.16 (Properties of g_T).
Let Assumption 5.15 be fulfilled. Then $g_T \in C(\mathbb{R}^K \times \mathcal{I} \times \mathbb{R}^K, \mathbb{R}^K)$ is Lipschitz continuous w.r.t. $A_T^\top e$ and θ. Furthermore the functions G_k are bounded and they are monotone decreasing functions, i.e.

$$(G_k(\theta_k) - G_k(\bar{\theta}_k))(\theta_k - \bar{\theta}_k) \le 0 \quad \forall \theta_k, \bar{\theta}_k \in \mathbb{R}.$$

PROOF:

The continuity of g_T is obvious. It is $\widetilde{G}_k(\theta_k) \leq \widetilde{G}_k(\beta_k)$ for all $\theta_k \geq \beta_k$ because $R_{0k}, \alpha_k > 0$. So the G_k are uniformly bounded, i.e. there is a $C_G > 0$ such that

$$\|G(\theta)\|_* \leq C_G \quad \forall \theta \in \mathbb{R}^K.$$

Hence g_T is Lipschitz continuous w.r.t. $A_T^\top e$. For proving the Lipschitz continuity of g_R w.r.t. θ we prove the Lipschitz continuity of G_k with a constant $L_G > 0$ being independent of k. Let $\theta_k, \bar{\theta}_k \in \mathbb{R}$. If $\theta_k, \bar{\theta}_k \geq \beta_k$ we see that

$$
\begin{aligned}
\left|G_k(\theta_k) - G_k(\bar{\theta}_k)\right| &= R_{0k}^{-1}\left|\frac{1}{1 + \alpha_k(\theta_k - \theta_{0k})} - \frac{1}{1 + \alpha_k(\bar{\theta}_k - \theta_{0k})}\right| \\
&= R_{0k}^{-1}\alpha_k \left|\theta_k - \bar{\theta}_k\right| G_k(\theta_k)G_k(\bar{\theta}_k) \\
&\leq L_G \left|\theta_k - \bar{\theta}_k\right|
\end{aligned}
$$

with

$$L_G = \max_{1 \leq k \leq K} R_{0k}^{-1}\alpha_k C_G^2$$

because G is bounded. With a similar argument we validate the Lipschitz continuity for the cases $\theta_k, \bar{\theta}_k < \beta_k$ and $\theta_k < \beta_k$, $\bar{\theta}_k \geq \beta_k$ with the same constant L_G. Hence g_T is Lipschitz continuous w.r.t. θ.

The monotonicity of G_k can be seen as follows. For $\beta_k \leq \theta_k \leq \bar{\theta}_k$ we have

$$\widetilde{G}_k(\bar{\theta}_k) \leq \widetilde{G}_k(\theta_k) \leq \widetilde{G}_k(\beta_k)$$

because $R_{0k}, \alpha_k > 0$. So G_k is monotone decreasing. □

As shown above the modification of $\widetilde{G}_k$ in Assumption 5.15 was necessary to prove the Lipschitz and monotonicity properties of G_k on $\mathbb{R}$. The function $\widetilde{G}_k$ itself is not Lipschitz continuous and has a pole at $-\alpha_k^{-1} + \theta_{0k}$. However, the model is only physically reasonable for temperatures in a neighborhood of the reference temperature θ_{0k}. Therefore the presented cutoff of the function $\widetilde{G}_k$ is not problematic, cf. [Bar04]. A similar cutoff has to be made for the power term W_k.

Assumption 5.17.

Let Assumption 5.15 be fulfilled. Let $\gamma > 0$ be given and assume the power terms W_k, $k = 1, \ldots, K$, to be of the form

$$W_k(\theta_k, y) := \begin{cases} G_k(\theta_k)y^2, & \text{if } |y| \leq \gamma, \\ G_k(\theta_k)(2\gamma |y| - \gamma^2), & \text{else.} \end{cases}$$

Let the operator $\mathcal{W}$ be given by (5.49).

First we prove a helpful lemma.

Lemma 5.18.
Let $C > 0$ and define $f : \mathbb{R} \to \mathbb{R}$ as

$$f(x) := \begin{cases} x^2, & \text{if } |x| \le C, \\ 2C\,|x| - C^2, & \text{if } |x| > C. \end{cases}$$

Then f is Lipschitz continuous.

PROOF:
Let $x_1, x_2 \in \mathbb{R}$. If $|x_1| \le C$, $|x_2| \le C$ we have

$$|f(x_1) - f(x_2)| = |x_1 + x_2|\,|x_1 - x_2| \le 2C\,|x_1 - x_2|\,.$$

If $|x_1| > C$, $|x_2| > C$ the function f is affin linear and clearly fulfills a Lipschitz condition with constant $2C > 0$.
If $|x_1| \le C$, $|x_2| > C$, we obtain using the inverse triangle inequality

$$\begin{aligned} |f(x_2) - f(x_1)| &= 2C\,|x_2| - C^2 - x_1^2 \\ &= 2C\,|x_2| - 2C^2 - (C^2 + x_1^2) \\ &\le 2C(|x_2| - |x_1|) \\ &\le 2C\,|x_1 - x_2| \end{aligned}$$

because $C^2 \ge C\,|x_1|$ and $(C^2 + x_1^2) > 0$. The case $|x_1| > C$, $|x_2| \le C$ follows analogously and hence we find a global Lipschitz bound $2C > 0$. □

Lemma 5.19 (Properties of $\mathcal{W}$).
Let Assumptions 5.15 and 5.17 be fulfilled. Then $\mathcal{W} \in C(V \times \mathbb{R}^{n_e}, V^*)$ is Lipschitz continuous w.r.t. e and monotone w.r.t. u.

PROOF:
For all $1 \le k \le K$ the functions G_k are continuous and so are the functions W_k. Because of Lemma 5.14 the operators $\mathcal{K}_k$ are Lipschitz continuous and so $\mathcal{W}$ is also continuous. The functions G_k are uniformly bounded because of Lemma 5.16 and so the functions W_k are Lipschitz continuous w.r.t. e. This is a consequence of Lemma 5.18. For $e, \bar{e} \in \mathbb{R}^{n_e}$, $u, v \in V$ we have

$$\langle \mathcal{W}(u, e) - \mathcal{W}(u, \bar{e}), v \rangle_V = -\sum_{k=1}^{K} \frac{1}{\alpha\,|\Omega_k|}(W_k(\mathcal{K}_k(u), A_{T,k}^\top e) - W_k(\mathcal{K}_k(u), A_{T,k}^\top \bar{e}))(\chi_k\,|\,v)_H$$

$$\le c_{\mathcal{W}}\,\|e - \bar{e}\|\,\|v\|_V$$

with $c_{\mathcal{W}} > 0$ because the embedding $V \subseteq H$ is continuous. This proves the Lipschitz continuity of $\mathcal{W}$ w.r.t. e. For checking the monotonicity of $\mathcal{W}$ w.r.t. u let $e \in \mathbb{R}^{n_e}$ and

$u, \bar{u} \in V$. For fixed applied potential the functions W_k are monotone decreasing because the G_k are, cf. Lemma 5.16. We set

$$\theta_k := \mathcal{K}_k(u), \quad \bar{\theta}_k := \mathcal{K}_k(\bar{u})$$

and we see with the definition of $\mathcal{K}_k$ from (5.47) that

$$\langle \mathcal{W}(u,e) - \mathcal{W}(\bar{u},e), u - \bar{u} \rangle_V$$

$$= -\sum_{k=1}^{K} \frac{1}{\alpha |\Omega_k|} (W_k(\mathcal{K}_k(u), A_{T,k}^{\top}e) - W_k(\mathcal{K}_k(\bar{u}), A_{T,k}^{\top}e))(\chi_k | u - \bar{u})_H$$

$$= -\sum_{k=1}^{K} \frac{1}{\alpha} (W_k(\theta_k, A_{T,k}^{\top}e) - W_k(\bar{\theta}_k, A_{T,k}^{\top}e))(\theta_k - \bar{\theta}_k) \geq 0.$$

The last line follows from Assumption 5.17 and Lemma 5.16. $\qquad\qquad\square$

Considering the prototype (5.19) we have to restrict ourselves to linear, but still time-dependent functions q_C and ϕ_L. Furthermore we have to exclude any temperature dependence from the algebraic part of the MNA equations. This is also required in [Bar04] in the 1D setting. Therefore we make the following assumptions.

Assumption 5.20.
The functions q_C and ϕ_L are linear, i.e.

$$q_C(v_C, t) := C(t)v_C, \quad \phi_L(j_L, t) := L(t)j_L$$

with matrices $C(t) \in \mathbb{R}^{n_C \times n_C}$ and $L(t) \in \mathbb{R}^{n_L \times n_L}$ for $t \in \mathcal{I}$.

Assumption 5.21.
Let Q_C be a projector onto $\ker A_C^{\top}$. Then we assume $A_T^{\top} Q_C = 0$.

Note that Assumption 5.20 in combination with Assumptions 3.21 and 3.22 ensures that q_C and ϕ_L are continuously differentiable and so the matrices $C(t)$ and $L(t)$ are continuously differentiable and positive definite for all $t \in \mathcal{I}$. We can now apply the solvability result from Theorem 5.11 to the system (5.50).

Theorem 5.22 (Unique solvability of (5.50)).
Let be $\mathcal{I} := [t_0, T]$ and $V := H_0^1(\Omega)$, $H := L_2(\Omega)$ with Ω being defined as before. Let Assumptions 3.21, 3.22, 3.26, 5.15, 5.17, 5.20 and 5.21 be fulfilled. Furthermore we assume

(i) $\mathcal{B}$ given by (5.45) is linear, strongly monotone and bounded.

(ii) The initial values $u(t_0) = u_0 \in H$, $(\mathrm{p}_C^{\top}e(t_0), j_L(t_0))^{\top} = (\mathrm{p}_C^{\top}e_0, j_{L0})^{\top}$ are given.

(iii) *Let $\{v_1, v_2, \dots\}$ be a basis of V in the sense of Definition A.11. Furthermore set $V_n := \{v_1, \dots, v_n\}$ and let there be a sequence $(u_{n0}) \subseteq V$ with $u_{n0} \in V_n$ and $u_{n0} \to u_0$ in H as $n \to \infty$.*

Then the system (5.50) has a unique solution $(e, j_L, j_V, u) \in C(\mathcal{I}, \mathbb{R}^{n_e+n_L+n_V} \times H)$ with $(\mathrm{p}_C^\top e, j_L) \in C^1(\mathcal{I}, \mathbb{R}^{k_C+n_L})$ and $u \in W_2^1(\mathcal{I}; V, H)$. Furthermore we have for the solution $(e_n, (j_L)_n, (j_V)_n, u_n)$ of the corresponding Galerkin equations that

$$\|e_n - e\|_\infty + \|(j_L)_n - j_L\|_\infty + \|(j_V)_n - j_V\|_\infty \to 0$$

$$\max_{t \in \mathcal{I}} \|u_n(t) - u(t)\|_H \to 0$$

$$\|u_n - u\|_{L_2(\mathcal{I}, V)} \to 0$$

as $n \to \infty$.

PROOF:

We perform the decoupling for the MNA equations, cf. Lemma 3.29. The condition $A_T^\top Q_C = 0$ ensures that there is no dependency of u in the algebraic part, see the decoupling in (3.19). With the same notation as in Lemma 3.29 we obtain:

$$\frac{\mathrm{d}}{\mathrm{d}t} m\left(\begin{pmatrix} e_C \\ j_L \end{pmatrix}, t\right) - f\left(\begin{pmatrix} e_C \\ j_L \end{pmatrix}, e_{\overline{CV}}, t\right) + \begin{pmatrix} \mathrm{p}_C^\top A_T g_T(A_T^\top \mathrm{p}_C e_C, t, \mathcal{K}(u)) \\ 0 \end{pmatrix} = 0$$

$$g\left(\begin{pmatrix} e_C \\ j_L \end{pmatrix}, e_{\overline{CV}}, t\right) = 0 \quad (5.54)$$

$$\begin{pmatrix} e_{\overline{C}V} \\ j_V \end{pmatrix} - h\left(\begin{pmatrix} e_C \\ j_L \end{pmatrix}, e_{\overline{CV}}, t\right) = 0$$

$$u' + \mathcal{B}u + \mathcal{W}(u, e) = 0$$

Due to the splitting (3.18) of the potentials e we write

$$z = \begin{pmatrix} x & y \end{pmatrix}^\top = \left((x_1, x_2) \ (y_1, y_2, y_3)\right)^\top = \left((e_C, j_L) \ (e_{\overline{CV}}, e_{\overline{C}V}, j_V)\right)^\top$$

and have

$$e = \mathrm{p}_C x_1 + \mathrm{q}_C \mathrm{q}_{CV} y_1 + \mathrm{q}_C \mathrm{p}_{CV} y_2.$$

With Assumptions 3.21 and 5.20 we have

$$\frac{\mathrm{d}}{\mathrm{d}t} m\left(\begin{pmatrix} e_C \\ j_L \end{pmatrix}, t\right) = \frac{\mathrm{d}}{\mathrm{d}t} \begin{pmatrix} \mathrm{p}_C^\top A_C q_C(A_C^\top \mathrm{p}_C e_C, t) \\ \phi_L(j_L, t) \end{pmatrix}$$

$$= \begin{pmatrix} \mathrm{p}_C^\top A_C C'(t) A_C^\top \mathrm{p}_C e_C + \mathrm{p}_C^\top A_C C(t) A_C^\top \mathrm{p}_C e_C' \\ L'(t) j_L + L(t) j_L' \end{pmatrix}$$

$$= M_{CL}'(t) \begin{pmatrix} e_C \\ j_L \end{pmatrix} + M_{CL}(t) \begin{pmatrix} e_C' \\ j_L' \end{pmatrix}$$

with the matrix

$$M_{CL}(t) := \begin{pmatrix} p_C^\top A_C C(t) A_C^\top p_C & 0 \\ 0 & L(t) \end{pmatrix}.$$

$M_{CL}(t)$ is positive definite because $C(t)$, $L(t)$ are positive definite and

$$\ker (p_C^\top A_C C(t) A_C^\top p_C) = \ker A_C^\top p_C = \{0\}.$$

The system (5.54) can now be equivalently formulated in the form (5.19) with

$$\bar{f}(z,u,t) := M_{CL}(t)^{-1} \left(M'_{CL}(t)x - f(x,y_1,t) + \begin{pmatrix} p_C^\top A_T g_T(A_T^\top p_C x_1, t, \mathcal{K}(u)) \\ 0 \end{pmatrix} \right)$$

$$\bar{g}(z,t) := \begin{pmatrix} g(x,y_1,t) \\ (y_2, y_3)^\top - h(x,y_1,t) \end{pmatrix}$$

$$\mathcal{R}(u,z,t) := \mathcal{W}(u, p_C x_1 + q_C q_{CV} y_1 + q_C p_{CV} y_2)$$

We obtain the system

$$\begin{aligned} x' + \bar{f}(z,u,t) &= 0, \quad t \in \mathcal{I}, \\ \bar{g}(z,t) &= 0, \\ u' + \mathcal{B}u + \mathcal{R}(u,z,t) &= 0, \quad \text{in } V^*, \\ x(t_0) &= x_0, \quad u(t_0) = u_0 \end{aligned}$$

In order to apply Theorem 5.11 we have to check Assumption 5.7. Here (i), (ii), (v) and (vii) are obvious.

(iii) As a combination of continuous functions the function $\bar{f}$ is also continuous. For the Lipschitz continuity of $\bar{f}$ w.r.t. z and u we remark that the map $w \rightarrow M_{CL}(t)w$ is strongly monotone due to Assumption 3.22. As a consequence the inverse map is Lipschitz continuous and thus there is a $C_M > 0$ such that

$$\left\| M_{CL}^{-1}(t) \right\|_* \leq C_M.$$

Since $M'_{CL}(t)$ itself is bounded and the functions f, g_T and $\mathcal{K}$ are Lipschitz continuous w.r.t. z and u, the Lipschitz continuity of $\bar{f}$ follows.

(iv) This can be seen in the proof of Theorem 3.30.

(vi) Due to Lemma 5.19 the operator $\mathcal{R}$ is continuous, Lipschitz continuous w.r.t. z and monotone w.r.t. u. The boundedness follows directly because

$$\mathcal{R}(u(t), 0, t) = \mathcal{W}(u(t), 0) = 0.$$

So Theorem 5.11 is applicable and gives the desired result for the decoupled system and hence also for the system (5.50). $\qquad\square$

We note that $\mathcal{B}$ as given in (5.45) is linear. If $\alpha, \hat{\alpha} > 0$ and $\kappa = (\kappa_{ij})_{i,j=1,\ldots,d} \in \mathbb{R}^{d \times d}$ is positive definite then $\mathcal{B}$ is also bounded and strongly monotone, cf. Lemma 5.4 (i). We give some additional remarks concerning the comparison to the analysis presented in [Cul09] and show that system (5.50) has Perturbation Index 1.

Remark 5.23 (Comparison to the analytical results in [Cul09]).
In [Cul09] system (5.44) is also treated by the Galerkin approach and a unique solvability result is stated, cf. [Cul09, Theorem 3.10]. Although this result seems to be of more generality than Theorem 5.22, it has to be dealt with carefully. We give a few reasons. In [Cul09] a priori estimates for the MNA equations have been assumed, but not been proven under the assumptions given for the (nonlinear) MNA equations. Furthermore for the unique solvability of the corresponding Galerkin equations it is referred to a global solvability theorem for DAEs having Tractability Index 1 ([GM86, Theorem 15]). The assumptions of this theorem have not been validated, especially the uniform boundedness of a certain inverse matrix is not proven. Additionally, the convergence proof of the Galerkin solutions ([Cul09, Theorem 3.10]) is not complete. It lacks an explanation how uniform convergence of the Galerkin solutions is achieved having only weak convergence in the beginning. This was a main difficulty in the prototype before (Theorems 5.11 and 5.22) and we had to exclude the temperature dependence from the algebraic part.

Remark 5.24 (Application of the perturbation result).
Let be $\mathcal{I} := [t_0, T]$ and $V := H_0^1(\Omega)$, $H := L_2(\Omega)$ with Ω being defined as before. Let Assumptions 3.21, 3.22, 3.26, 5.15, 5.17, 5.20 and 5.21 be fulfilled. Then we consider the perturbed system

$$
\begin{aligned}
A_C \frac{\mathrm{d}}{\mathrm{d}t} q_C(A_C^\top e, t) + A_R g_R(A_R^\top e, t) + A_T g_T(A_T^\top e, t, \mathcal{K}(u)) & \\
+ A_L j_L + A_V j_V + A_I i_s(t) &= \delta_e(t) \\
\frac{\mathrm{d}}{\mathrm{d}t} \phi_L(j_L, t) - A_L^\top e &= \delta_L(t) \\
A_V^\top e - v_s(t) &= \delta_V(t) \\
u' + \mathcal{B}u + \mathcal{W}(u, e) &= \delta_u(t)
\end{aligned}
\tag{5.55}
$$

for perturbations $\delta_e \in C(\mathcal{I}, \mathbb{R}^{n_e})$, $\delta_L \in C(\mathcal{I}, \mathbb{R}^{n_L})$, $\delta_V \in C(\mathcal{I}, \mathbb{R}^{n_V})$ and $\delta_u \in C(\mathcal{I}, V^*)$. Furthermore we assume

(i) $\mathcal{B}$ given by (5.45) is linear, strongly monotone and bounded.

(ii) The initial values $u(t_0) = u_0, u_0^\delta \in H$ and

$$
x(t_0) := (p_C^\top e(t_0), j_L(t_0))^\top = (p_C^\top e_0, j_{L0})^\top =: x_0, \quad x(t_0) = (p_C^\top e_0^\delta, j_{L0}^\delta)^\top =: x_0^\delta
$$

are given.

(iii) Let $\{v_1, v_2, \dots\}$ be a basis of V in the sense of Definition A.11. Furthermore set $V_n := \operatorname{span}\{v_1, \dots, v_n\}$ and let there be two sequences $(u_{n0}), (u_{n0}^\delta) \subseteq V$ with $u_{n0}, u_{n0}^\delta \in V_n$ and

$$
u_{n0} \to u_0 \text{ in } H, \quad u_{n0}^\delta \to u_0^\delta \text{ in } H \quad \text{as } n \to \infty.
$$

Then we see with the arguments in Theorem 5.22 and with Theorem 5.13 that (5.55) has a unique solution $(e^\delta, j_L^\delta, j_V^\delta, u^\delta) \in C(\mathcal{I}, \mathbb{R}^{n_e+n_L+n_V} \times H)$ with $(\mathrm{p}_C^\top e^\delta, j_L^\delta) \in C^1(\mathcal{I}, \mathbb{R}^{k_C+n_L})$ and $u^\delta \in W_2^1(\mathcal{I}; V, H)$ for the initial values $x_0^\delta \in \mathbb{R}^{n_x}$ and $u_0^\delta \in H$. Furthermore we have the perturbation estimate

$$\left\| e - e^\delta \right\|_\infty + \left\| j_L - j_L^\delta \right\|_\infty + \left\| j_V - j_V^\delta \right\|_\infty + \max_{t \in \mathcal{I}} \left\| u(t) - u^\delta(t) \right\|_H + \left\| u - u^\delta \right\|_{L_2(\mathcal{I},V)}$$

$$\leq \; C \left(\left\| x_0 - x_0^\delta \right\| + \left\| u_0 - u_0^\delta \right\|_H + \left\| \delta_e \right\|_\infty + \left\| \delta_L \right\|_\infty + \left\| \delta_V \right\|_\infty + \max_{t \in \mathcal{I}} \left\| \delta_u(t) \right\|_{V^*} \right)$$

for a $C > 0$ where (e, j_L, j_V, u) is the solution to (5.50) with initial values $x_0 \in \mathbb{R}^{n_x}$ and $u_0 \in H$. Hence system (5.50) has Perturbation Index 1.

5.5. Conclusion

This chapter has been dedicated to the derivation of two prototypes of coupled systems as given in (5.1) and (5.19) respectively. In the first one (elliptic prototype) a semilinear finite dimensional differential-algebraic equation (DAE) is coupled to an infinite dimensional algebraic operator equation. The right hand side of the operator equation depends on the DAE variable z and the dynamical part of the DAE depends on the variable u of the operator equation. So the coupling is two-directional. For the second prototype (parabolic prototype) the DAE and the coupling are similar to the first one. But in the second case the DAE is coupled to an evolution equation. We studied both systems with regard to unique solvability and strong convergence of solutions of the corresponding Galerkin equations. In the elliptic case solvability was shown independently of the Galerkin approach (Theorem 5.1) and the convergence of the Galerkin solutions can be treated separately, cf. Theorem 5.3. In the parabolic case a unique solution is found as the limit of a sequence of Galerkin solutions, cf. Theorem 5.11. In both cases we obtain global solvability results which are mainly based on the ideas of chapter 3, the theory of monotone operators in Banach spaces, see e.g. [Zei90b], and the Theorem of Arzelà-Ascoli. Furthermore we showed that both system have Perturbation Index 1, cf. Theorems 5.2 and 5.13.

We also presented two coupled systems to which the prototype systems could be applied. Both systems are examples for coupled systems in circuit simulation involving the equations of the Modified Nodal Analysis (MNA). In the elliptic case we coupled the MNA equations to the Laplace equation for the electrostatic potential which is derived by Ohm's law. The node potentials serve as boundary conditions for the Laplace equation and a term for the current also influences the MNA, cf. the classical formulation (5.12) and the corresponding weak formulation (5.16). In the parabolic case we relied on a model by [Cul09] which adds thermal effects to the MNA equations. Thus the heat equation is coupled to the MNA equations, cf. systems (5.44) and (5.50). On the one hand thermally active resistors depend nonlinearly on the temperature whereas on the other hand certain power terms depending on the node potentials have an influence on the right hand side of the heat equation. In both the elliptic and the parabolic case the results of the corresponding prototype systems could be applied under suitable topological conditions to the circuit, cf. Remark 5.5, Theorem 5.22 and Remark 5.24.

The prototype systems presented in this chapter are a first step towards a systematic treatment of coupled systems. We see many ways of extending the results in this chapter. A main task for future work could be to allow the algebraic part of the DAE to depend on u as well. Appropriate conditions have to be found to ensure the solvability of the algebraic part. Furthermore, considering the Galerkin approach, it is not yet clear how in this case the strong convergence of the Galerkin solutions can be shown having only weak convergence in the beginning, compare here the proofs of Theorems 5.3 and 5.11. A further important step of extending the results could be to investigate the higher index case. In literature solvability results for specific coupled systems in circuit simulation are all restricted to the topological index 1 case, see [Gün01, ABGT03, Bar04, ABG10]. From a numerical point of view it would be also very interesting to obtain an error estimate for the Galerkin equations. With such an error estimate the error of the completely discretized system (in space and time) with regard to the original system could be obtained. The only result in this perspective can be found in [MT11] for a system coupling the MNA equations to the stationary drift diffusion equations in the index 1 setting.

6. Conclusion and Outlook

This thesis is devoted to the study of so-called abstract differential-algebraic equations (ADAEs). They arise in many application fields, for example in circuit simulation where they are also known as coupled systems. Therefore we presented a general framework for these kind of equations and discussed their treatment in the literature. We proposed guiding questions to be investigated when studying nonlinear ADAEs. They are concerned with unique solvability, perturbation behavior and convergence of solutions of discretization methods of ADAEs. We investigated nonlinear differential-algebraic equation (DAEs) with monotonicity properties and presented two approaches for treating ADAEs with regard to the guiding questions.

We have derived a global and a local existence result for a certain class of nonlinear differential-algebraic equations. The results are based on a decoupling using orthonormal bases of certain subspaces and monotonicity properties of the functions involved. Furthermore we have shown that these specific DAEs have Perturbation Index 1. The solvability results are applicable in circuit simulation to the equations of the Modified Nodal Analysis (MNA) under the topological index 1 conditions. So we obtained a first global solvability result for the MNA equations. Therefore the usual passivity assumptions for the element functions had to be slightly extended to match the concept of strong monotonicity. However, differentiability of the conductivity function is not necessary anymore. We also proved in this case that the MNA equations with the topological index 1 conditions still have Perturbation Index 1.

Additionally we studied an abstract approach for solving ADAEs with monotone operators on Banach spaces. In [Tis04] the theory of linear evolution equations with monotone operators was extended to a certain class of linear ADAEs. We extended this approach to the nonlinear case. In this setting we proved unique solvability by approximating the original system with the Galerkin approach. The solution is obtained as a weak limit of solutions to the Galerkin equations which yield a nonlinear DAE. We have shown global solvability of the Galerkin equations in a non-standard way because common solvability results from literature could not be applied due to lacking smoothness properties. We also showed strong convergence of the Galerkin solutions and a perturbation estimate yielding the investigated ADAE to have Perturbation Index 1. For suitable smoothness conditions we have also proven that the ADAE Index does not exceed one. As for the linear case the most interesting point is that the choice of the basis functions for the Galkerin approach is not comletely arbitrary. A proper choice ensures that the Galerkin solutions fulfill the constraints of the system.

Furthermore we presented two prototypes of coupled systems. In the first one (elliptic prototype) a semi-linear finite dimensional DAE is coupled to an infinite dimensional algebraic operator equation whereas in the second prototype (parabolic prototype) the DAE is coupled to an evolution equation. In both cases the coupling is two-directional. For both systems we proved existence and uniqueness of solutions and the strong convergence of solutions of the corresponding Galerkin equations. Furthermore it is shown that both systems have Perturbation Index 1. The results for the prototypes have been applied to exemplary coupled systems in circuit simulation. In the elliptic case the MNA equations are coupled to the Laplace equation simulating a specific resistor and in the parabolic case the MNA equations are coupled to the heat equation adding thermal effects of resistors to the circuit.

We see many directions for future research especially for coupled systems. The presented prototypes mark a first step towards a systematic treatment of coupled systems. Nevertheless the structure of the prototypes has to be extended, e.g. an occurence of the infinite dimensional variable in the algebraic part of the semi-explicit DAE would be desirable. Especially a convergence proof of the Galerkin solutions is a challenge here. Furthermore, for the exemplary coupled systems in circuit simulation we restricted ourselves to the MNA equations satisfying the topological index 1 conditions. A prototype which also covers the index 2 case is of interest. This would be a big step for a better understanding of coupled systems as e.g. solvability results for specific coupled systems in literature are all restricted to the topological index 1 case.

From a numerical point of view it would be also very interesting to obtain error estimates for the Galerkin equations of coupled systems. With such an error estimate the distance of the completely discretized system (in space and time) with regard to the original system could be measured.

A. Appendix

A.1. Projectors

Important properties of projectors taken from [Zei86] are summarized. In the following let X be a Banach space and V, W linear subspaces of X.

Definition A.1 (Projectors).

(i) A linear continuous operator $\mathcal{P} : X \rightarrow X$ is called a projector or projection operator if and only if $\mathcal{P}$ is idempotent, i.e. $\mathcal{P}^2 = \mathcal{P}$.

(ii) A projector $\mathcal{P}$ projects onto a linear subspace V of X if and only if $\operatorname{im} \mathcal{P} = V$.

(iii) A projector $\mathcal{P}$ projects along a linear subspace V of X if and only if $\ker \mathcal{P} = V$.

(iv) A projector $\mathcal{P} : \mathbb{R}^n \rightarrow \mathbb{R}^n$ is called orthogonal if $\mathcal{P}^\top = \mathcal{P}$.

(v) The sum $X = V \oplus W$ is called a direct sum if and only if for all $x \in X$ we have $x = v + w$ with a unique $v \in V$ and a unique $w \in W$.

(vi) The direct sum $X = V \oplus W$ is called a topological direct sum if and only if the operators $\mathcal{P}$ and $\mathcal{Q}$ defined as

$$\mathcal{P}x = v \quad \text{and} \quad \mathcal{Q}x = w \text{ for } \quad x = v + w, \quad v \in V, w \in W$$

are continuous. This means $\mathcal{P}$ and $\mathcal{Q}$ are projection operators.

Starting from these definitions we can list some properties.

- If $\mathcal{P} : X \rightarrow X$ is a projector and $\operatorname{id} : X \rightarrow X$ is the identity map then $\mathcal{Q} := \operatorname{id} - \mathcal{P}$ is also a projector. $\mathcal{Q}$ is called the complementary projector of $\mathcal{P}$. From the fact that

$$\operatorname{im} \mathcal{P} = \ker \mathcal{Q} \quad \text{and} \quad \ker \mathcal{P} = \operatorname{im} \mathcal{Q}$$

we conclude that

  - if $\mathcal{P}$ projects onto V then $\mathcal{Q}$ projects along V and
  - if $\mathcal{P}$ projects along V then $\mathcal{Q}$ projects onto V.

- If $X = V \oplus W$ is a topological sum then V and W are closed. Conversely, if $X = V \oplus W$ is a direct sum and V and W are closed linear subspaces of the Banach space X, then this is a topological sum.

- If $X = V \oplus W$ is a topological sum we say that V splits the space X. This is equivalent to the condition that there exists a projector $\mathcal{P}$ onto V in which case $W = (\mathrm{id} - \mathcal{P})(X)$. Hence we have $X = \mathrm{im}\,\mathcal{P} \oplus \ker \mathcal{P}$. Furthermore every projector has a closed nullspace and a closed image space.

A.2. Theorem of Carathéodory

In this section we will state the well-known Theorem of Carathéodory. The results presented are taken from [Zei90b]. For given $x_0 \in \mathbb{R}^n$, $n \geq 1$ and $t_0 \in \mathbb{R}$ we consider the initial value problem

$$x'(t) = f(x(t), t), \quad t \in \mathcal{I} \tag{A.1a}$$
$$x(t_0) = x_0 \tag{A.1b}$$

with $f : \mathbb{R}^n \times \mathcal{I} \to \mathbb{R}^n$, $n \geq 1$, $\mathcal{I} \subseteq \mathbb{R}$ a closed interval and $t_0 \in \mathcal{I}$ and $x_0 \in \mathbb{R}^n$. Along with (A.1) we consider the integral equation

$$x(t) = x_0 + \int_{t_0}^{t} f(x(s), s)\mathrm{d}s, \quad t \in \mathcal{I}. \tag{A.2}$$

Furthermore we write $x = (x_1, \ldots, x_n)^\top$, $x_0 = (x_{01}, \ldots, x_{0n})^\top$ and $f = (f_1, \ldots, f_n)^\top$. So we can write problem (A.1) as follows:

$$x_i'(t) = f_i(x(t), t), \quad i = 1, \ldots, n, \quad t \in \mathcal{I},$$
$$x_i(t_0) = x_{i0}.$$

Theorem A.2 (Carathéodory).
Let $\mathcal{I} \subseteq J$ and set

$$J := \{t \in \mathbb{R}|\ |t - t_0| \leq r_0\},$$
$$K := \{x \in \mathbb{R}^n|\ \|x - x_0\| \leq r\}$$

where $r, r_0 > 0$. We assume that $f : K \times J \to \mathbb{R}$ satisfies the Carathéodory conditions, i.e. for all $i \in \{1, \ldots, n\}$:

$$x \mapsto f_i(x, t) \text{ is continuous on } K \text{ for almost all } t \in J, \tag{A.3a}$$
$$t \mapsto f_i(x, t) \text{ is measurable on } J \text{ for each } x \in K. \tag{A.3b}$$

Furthermore let there be an integrable function $M : J \to \mathbb{R}$ such that

$$|f_i(x, t)| \leq M(t) \text{ for all } (x, t) \in K \times J \text{ and all } i. \tag{A.4}$$

Then we have the following:

(i) *There exists an open neighborhood U of t_0 and a continuous function $x(\cdot) : U \to \mathbb{R}^n$ which solves the integral equation* (A.2).

(ii) *For almost all $t \in U$ the derivative $x'(t)$ exists and* (A.1) *holds.*

(iii) *The components $x_1(\cdot), \ldots, x_n(\cdot)$ of $x(\cdot)$ have generalized derivatives on U and $x(\cdot)$ is a solution of* (A.1) *on U in the sense of generalized derivatives. $x(\cdot)$ is also called a solution of* (A.1) *in the sense of Carathéodory.*

Theorem A.2 remains true if J is a one-sided neighborhood of t_0, i.e.

$$J = \{t \in \mathbb{R} \mid 0 \le t - t_0 \le r_0\}.$$

Then $U = \{t \in \mathbb{R} \mid 0 \le t - t_0 \le r_1\}$ with $r_1 > 0$. A proof of Theorem A.2 can be found in [CL55, p.43] or [Kam60, p. 197]. If the assumptions are satisfied for $K = \mathbb{R}^n$ then $U = J$. If a priori estimates of a possible solution are known it can be extended to the boundary. This is stated in the next theorem which is an application of Theorem A.2.

Theorem A.3 (Continuation of solutions).
Suppose there exists a $C > 0$ such that for any solution $x(\cdot) : U \to \mathbb{R}^n$ of (A.1) *on any arbitrary subinterval $U \subseteq \mathcal{I}$ the estimate $\|x(t)\| \le C$ holds on U. Let $J := \mathcal{I}$ and $K := \{x \in \mathbb{R}^n : \|x\| \le 2C\}$ and let the Carathéodory conditions* (A.3) *and the growth condition* (A.4) *be fulfilled on $K \times J$. Then* (A.1) *has a (global) solution in the sense of Carathéodory on $\mathcal{I}$.*

A proof of this statement is given in [Zei90b, P.30.2].

A.3. Basics for Evolution Equations

In this section we summarize briefly basic spaces and properties which are important in the context of evolution equations. For a deeper treatment we refer to standard literature, e.g. [Zei90a] or [Emm04].

Evolution triples

Definition A.4.
Let

(i) Z be a real, separable and reflexive Banach space with dual space Z^*,

(ii) H be a real, separable Hilbert space,

(iii) Z be dense in H and the embedding $Z \subseteq H$ be continuous, i.e. there is a constant $c > 0$ such that

$$\|z\|_H \le c \|z\|_Z \quad \forall z \in Z.$$

Then the spaces Z, H and Z^* form a so-called evolution triple or Gelfand triple. We write

$$Z \subseteq H \subseteq Z^*.$$

A Banach space Z is called separable if it has a countable dense subset. A subset of a separable space is separable itself. Note that a Hilbert space is separable if and only if it has a countable orthonormal basis. Z is called reflexive if the natural embedding $i_Z : Z \to Z^{**}$ is surjective. Here $Z^{**} = (Z^*)^*$ is the double dual space and i_Z is given by

$$\langle i_Z z, z^* \rangle_{Z^*} := \langle z^*, z \rangle_Z, \quad \forall z \in Z, \ z^* \in Z^*.$$

The map i_Z is linear, continuous and isometric, i.e. $\|i_Z z\|_{Z^{**}} = \|z\|_Z$ for all $z \in Z$. So it is bijective on the reflexive space Z and thus $Z = Z^{**}$ in the isomorphic sense. We can identify z with $i_Z z$ and write

$$\langle z^*, z \rangle_{Z^*} = \langle z, z^* \rangle_Z \quad \forall z \in Z, \ z^* \in Z^*.$$

Theorem A.5 (Riesz Representation Theorem).
Let H be a real Hilbert space. Then the map $i_H : H \to H^*$, $y \mapsto (\cdot\,|\,y)_H$ is linear, isometric and bijective. In other words, for $y^* \in H^*$ there exists a unique $y \in H$ with $y^*(x) = (x\,|\,y)_H$ for all $x \in H$ and $\|y^*\|_{H^*} = \|y\|_H$.

For a proof of this famous result we refer to [Zei90a, chapter 18.11b] or [Wer05, chapter V.3]. With the map i_H we see that $H = H^*$ in the isomorphic sense and we can write $\langle y, x \rangle_H = (x\,|\,y)_H$ for all $x, y \in H$. As a direct consequence every Hilbert space is reflexive. A similar version of Theorem A.5 holds for evolution triples and it explains how the inclusion $H \subseteq Z^*$ has to be understood.

Proposition A.6.
Let $Z \subseteq H \subseteq Z^*$ be an evolution triple. Then the following holds.

(i) To each $u \in H$ there corresponds an element $\bar{u} \in Z^*$ which is defined by

$$\langle \bar{u}, z \rangle := (u\,|\,z)_H \quad \forall z \in Z.$$

(ii) The mapping $j : u \mapsto \bar{u}$ from H to Z^* is linear, injective and continuous.

PROOF:
(i) Given $u \in H$, then $\bar{u}$ as defined above is a linear functional on Z. Furthermore we have

$$\langle \bar{u}, z \rangle = (u\,|\,z)_Z \leq \|u\|_H \|z\|_H \leq c \|u\|_H \|z\|_Z \quad \forall z \in Z$$

because the embedding $Z \subseteq H$ is continuous. This implies the continuity of $\bar{u}$ because $\|\bar{u}\|_{Z^*} \leq c \|u\|_H$ and thus $\bar{u} \in Z^*$.

(ii) The mapping $j : H \rightarrow Z^*$ is linear and continuous because of (i). For injectivity we assume $j(u) = \bar{u} = 0$. This implies

$$(u \,|\, z)_H = 0 \quad \forall z \in Z.$$

Since Z is dense in H we have $(u \,|\, z)_H = 0$ for all $z \in H$ and get $u = 0$. $\qquad\square$

Proposition A.6 allows us to identify $u \in H$ with $\bar{u} \in Z^*$ and so we write

$$\langle u, z \rangle_Z = (u \,|\, z)_H \quad \forall u \in H, \ z \in Z,$$
$$\|u\|_{Z^*} \leq c \|u\|_H \quad \forall u \in H.$$

The embeddings $Z \subseteq H$ and $H = H^* \subseteq Z^*$ are both continuous. In particular we have

$$\langle z, w \rangle_Z = \langle w, z \rangle_Z \quad \forall z, w \in Z$$

because $(w \,|\, z)_H = (z \,|\, w)_H$.

The Lebesgue spaces $L_p(\mathcal{I}, V)$

Let V be a real Banach space and $\mathcal{I} := [t_0, T] \subseteq \mathbb{R}$ with $t_0 < T < \infty$. The following results are taken from [Zei90a, chapter 23] where also more properties can be found. We state them here for $1 \leq p < \infty$. In this thesis we mainly consider the case $p = 2$. The space $L_p(\mathcal{I}, V)$ consists of all measurable functions $u : \mathcal{I} \rightarrow V$ satisfying

$$\|u\|_{L_p(\mathcal{I},V)} := \left(\int_{\mathcal{I}} \|u(t)\|_V^p \, dt \right)^{\frac{1}{p}} < \infty.$$

With the norm $\|\cdot\|_{L_p(\mathcal{I},V)}$ the space $L_p(\mathcal{I}, V)$ is a real Banach space. The following two results can be found in [Zei90a, Propositions 23.6 and 23.7].

Proposition A.7 (Hölder inequality).
Let V be a Banach space. Then

$$\int_{\mathcal{I}} |\langle v(t), u(t) \rangle_V| \, dt \leq \left(\int_{\mathcal{I}} \|v(t)\|_{V^*}^q \, dt \right)^{\frac{1}{q}} \left(\int_{\mathcal{I}} \|u(t)\|_V^p \, dt \right)^{\frac{1}{p}}$$

holds for all $u \in L_p(\mathcal{I}; V)$, $v \in L_q(\mathcal{I}; V^*)$ with $1 < p < \infty$, $p^{-1} + q^{-1} = 1$. In particular all the integrals above exist.

Proposition A.8.
Let V be a real, reflexive and separable Banach space and let $1 < p < \infty$, $p^{-1} + q^{-1} = 1$. Then the following holds:

(a) The map $j : L_q(\mathcal{I}, V^*) \to L_p(\mathcal{I}, V)^*$, $j(v) = \bar{v}$ with

$$\langle \bar{v}, u \rangle_{L_p(\mathcal{I}, V)^*} := \int_{\mathcal{I}} \langle v(t), u(t) \rangle_V \mathrm{d}t \quad \forall u \in L_p(\mathcal{I}; V)^*$$

is well-defined, linear, bijective and isometric, i.e. $\|\bar{v}\|_{L_p(\mathcal{I}, V)^*} = \|v\|_{L_q(\mathcal{I}, V^*)}$.

(b) The Banach space $L_p(\mathcal{I}, V)$ is reflexive and separable.

Proposition A.8(a) allows to identify $v \in L_q(\mathcal{I}, V^*)$ with $\bar{v} \in L_p(\mathcal{I}, V)^*$ and hence it is $L_p(\mathcal{I}, V)^* = L_q(\mathcal{I}, V^*)$ and we write for all $u, v \in L_p(\mathcal{I}, V)$:

$$\langle v, u \rangle_{L_p(\mathcal{I}, V)} = \int_{\mathcal{I}} \langle v(t), u(t) \rangle_V \mathrm{d}t,$$

$$\|v\|_{L_q(\mathcal{I}, V^*)} = \left(\int_{\mathcal{I}} \|v(t)\|_{V^*}^q \, \mathrm{d}t \right)^{\frac{1}{q}}.$$

Generalized derivatives

Generalized derivatives for functions with values in Banach spaces can be defined similar to weak derivatives of real-valued functions. We point out the basic definition and its basic treatment in the context of evolution triples. For an overview on useful results concerning generalized derivatives we refer to [Zei90a, chapter 23.5]. Let Z and Y be Banach spaces and $\mathcal{I} := [t_0, T] \subseteq \mathbb{R}$ with $t_0 < T < \infty$. Furthermore let $u \in L_1(\mathcal{I}, Z)$ and $w \in L_1(\mathcal{I}, Y)$. Then the function w is called the generalized derivative of the function u on $\mathcal{I}$ if and only if

$$\int_{\mathcal{I}} \varphi'(t) u(t) \mathrm{d}t = - \int_{\mathcal{I}} \varphi(t) w(t) \mathrm{d}t \quad \forall \varphi \in C_0^\infty(\mathcal{I}). \tag{A.5}$$

We write u' for w. The values of the functions $u : \mathcal{I} \to Z$ and $w : \mathcal{I} \to Y$ may lie in different spaces Z and Y. The equation above therefore includes the requirement that the integrals on both sides belong to $Z \cap Y$. The generalized derivative u' is unique. Considering the situation for evolution triples the following result holds.

Proposition A.9.
Let $Z \subseteq H \subseteq Z^*$ be an evolution triple, $\mathcal{I} := [t_0, T] \subseteq \mathbb{R}$ and let $1 < p < \infty$ with $p^{-1} + q^{-1} = 1$. Then the following holds:

(i) For $u \in L_p(\mathcal{I}, Z)$ the generalized derivative u' is unique as an element of $L_q(\mathcal{I}, Z^*)$, i.e. $t \mapsto u'(t)$ can be modified only on a subset of $\mathcal{I}$ of measure zero.

(ii) Let $u \in L_p(\mathcal{I}, Z)$. Then there exists $u' \in L_q(\mathcal{I}, Z^*)$ if and only if there exists $w \in L_q(\mathcal{I}, Z^*)$ such that

$$\int_{\mathcal{I}} (u(t) | z)_H \varphi'(t) \mathrm{d}t = - \int_{\mathcal{I}} \langle w(t), z \rangle_Z \varphi(t) \mathrm{d}t \quad \forall z \in Z \forall \varphi \in C_0^\infty(\mathcal{I})$$

Then $u' = w$ and

$$\frac{\mathrm{d}}{\mathrm{d}t}(u(t)\,|\,z)_H = \langle u'(t), z\rangle_Z$$

holds for all $z \in Z$ and almost all $t \in \mathcal{I}$. With $\frac{\mathrm{d}}{\mathrm{d}t}$ we denoted here the generalized derivative of real functions on $\mathcal{I}$.

In the situation of Proposition A.9 we define the space

$$W_p^1(\mathcal{I}; Z, H) := \{u \in L_p(\mathcal{I}, Z)\,|\, u' \in L_q(\mathcal{I}, Z^*)\}$$

which forms a real Banach space with the norm

$$\|u\|_{W_p^1} := \|u\|_{L_p(\mathcal{I}, Z)} + \|u'\|_{L_q(\mathcal{I}, Z^*)}\,.$$

Proposition A.10.
Let $Z \subseteq H \subseteq Z^*$ be an evolution triple, $\mathcal{I} := [t_0, T] \subseteq \mathbb{R}$ and let $1 < p < \infty$ with $p^{-1} + q^{-1} = 1$. Then it holds

(i) The embedding

$$W_p^1(\mathcal{I}; Z, H) \subseteq C(\mathcal{I}, H)$$

is continuous. More precisely, if $u \in W_p^1(\mathcal{I}; Z, H)$ then there exists a uniquely determined continuous function $\widetilde{u} : \mathcal{I} \to H$ which coincides almost everywhere on $\mathcal{I}$ with the initial function u. Henceforth we write u instead of $\widetilde{u}$. In this sense there is a $c > 0$ such that

$$\max_{t \in \mathcal{I}} \|u(t)\|_H \leq c \|u\|_{W_p^1}\,.$$

(ii) The set of all polynomials $w : \mathcal{I} \to Z$, i.e.

$$w(t) = \sum_i a_i t^i \quad \text{with } a_i \in Z, \ \forall i,$$

is dense in the spaces $W_p^1(\mathcal{I}; Z, H)$, $L_p(\mathcal{I}, Z)$ and $L_p(\mathcal{I}, H)$.

(iii) For all $u, z \in W_p^1(\mathcal{I}; Z, H)$ and arbitrary s, t with $t_0 \leq s \leq t \leq T$ the following integration by parts formula holds:

$$(u(t)\,|\,z(t))_H - (u(s)\,|\,z(s))_H = \int_s^t \langle u'(\tau), z(\tau)\rangle_Z + \langle z'(\tau), u(\tau)\rangle_Z \mathrm{d}\tau$$

Here the values $u(t), z(t), u(s), z(s)$ are the values of the corresponding continuous functions $u, z : \mathcal{I} \to H$ in the sense of (i).

Galerkin Schemes

Let V be a Banach space. We set

$$\text{dist}\,(u, Y) = \inf_{v \in Y} \|u - v\|_V$$

which is the minimal distance between the point $u \in V$ and the set $Y \subseteq V$.

Definition A.11.
A Galerkin Scheme in V is a sequence (Y_n) of finite dimensional nonzero subspaces $Y_n \subseteq V$ with

$$\lim_{n \to \infty} \text{dist}\,(u, Y_n) = 0 \quad \forall u \in V.$$

A basis of V is an at most countable sequence (w_j) of elements $w_j \in V$ where finitely many $w_1, \ldots, w_n$ are always linearly independent and

$$V = \overline{\bigcup_{n \in \mathbb{N}} V_n}$$

with $V_n = \text{span}\,\{w_1, \ldots, w_n\}$.

The following result can be found in [Zei90a, Proposition 21.49].

Proposition A.12. *(Existence of bases and Galerkin schemes)*
Let V be a separable Banach space. Then

(i) *V has a basis.*

(ii) *If (w_n) is a basis in V, then (V_n) with $V_n := \text{span}\,\{w_1, \ldots, w_n\}$ is a Galerkin Scheme in V.*

(iii) *If (Y_n) is a Galerkin Scheme in V, then we can construct a basis in V by means of (Y_n).*

More details and examples concerning Galerkin schemes are given in [Zei90a, chapter 21.13] or [Emm04].

A.4. Theorem of Browder-Minty

The following is taken from [Zei90b, chapter 26]. Let V be a real Banach space and consider the operator equation

$$\mathcal{B}(u) = f, \quad u \in V \tag{A.6}$$

with $\mathcal{B} : V \to V^*$ and $f \in V^*$. V^* is the dual space of V. We make the following definitions.

Definition A.13.
The operator $\mathcal{B} : V \to V^*$ is said to be

(i) strongly monotone if there is $\mu > 0$ such that

$$\langle \mathcal{B}(u) - \mathcal{B}(v), u - v \rangle_V \geq \mu \, \|u - v\|_V^2 \quad \forall u, v \in V.$$

If

$$\langle \mathcal{B}(u) - \mathcal{B}(v), u - v \rangle_V \geq 0 \quad \forall u, v \in V$$

then $\mathcal{B}$ is monotone.

(ii) hemicontinuous if the real function

$$t \mapsto \langle \mathcal{B}(u + tv), w \rangle_V$$

is continuous on $[0, 1]$ for all $u, v, w \in V$.

Having clarified these definitions we can state the following theorem. Note that in [Zei90b, Theorem 26.A] a more general version is presented and that we state it here tailored to our purposes.

Theorem A.14 (Browder-Minty).
Let V be a real reflexive Banach space. If $\mathcal{B} : V \to V^*$ is strongly monotone and hemicontinuous then (A.6) is uniquely solvable for all $f \in V^*$. Hence the inverse operator $\mathcal{B}^{-1} : V^* \to V$ exists and is Lipschitz continuous.

We also present a helpful lemma which can be found in [GGZ74, chaper 3, Lemma 1.3].

Lemma A.15.
Let V, W be real Banach spaces and $\mathcal{L} : W \to V$ be a linear operator with

$$\|\mathcal{L}u\|_V = \|u\|_W \quad \forall u \in W.$$

Let $\mathcal{L}^* : V^* \to W^*$ be the dual operator of $\mathcal{L}$ and $\mathcal{B} : V \to V^*$ be a possibly nonlinear operator. Then the operator $\overline{\mathcal{B}} := \mathcal{L}^*\mathcal{B}\mathcal{L} : W \to W^*$ is strongly monotone (hemicontinuous) if the operator $\mathcal{B}$ is strongly monotone (hemicontinuous).

PROOF:
This is an immediate consequence of the fact that $\mathcal{L}$ is linear, fulfills $\|\mathcal{L}u\|_V = \|u\|_W$ and that we have

$$\langle \overline{\mathcal{B}}u, w \rangle_W = \langle \mathcal{L}^*\mathcal{B}\mathcal{L}u, w \rangle_W = \langle \mathcal{B}\mathcal{L}u, \mathcal{L}w \rangle_V \quad \forall u, w \in W.$$

$\square$

A.5. Theorem of Arzelà-Ascoli

In this section the well-known Theorem of Arzelà-Ascoli is stated. We remind briefly that a subset M of a metric space V is relatively compact if the closure $\overline{M}$ is compact. This is exactly the case if every sequence in M has a convergent subsequence in V. The following can be found in [Die85, chapter 7] or [Zei90a, p.189]. Let $\mathcal{I} := [t_0, T]$ be a compact interval and let V be a Banach space.

Definition A.16 (Equicontinuity).
Let

$$H \subseteq \left\{ f : \mathcal{I} \to V \mid \sup_{t \in \mathcal{I}} \|f(t)\|_V < \infty \right\}.$$

Then the set H is equicontinuous at a point $\bar{t} \in \mathcal{I}$ if

$$\forall \varepsilon > 0 \; \exists \delta(\varepsilon) > 0 \; \forall f \in H \; \forall t \in \mathcal{I} : |t - \bar{t}| < \delta \Rightarrow \|f(t) - f(\bar{t})\|_V < \varepsilon.$$

The set H itself is called equicontinuous (on $\mathcal{I}$) if H is equicontinuous at every point $\bar{t}$ of $\mathcal{I}$.

Definition A.17 (Uniform convergence).
Let (f_n) be a sequence of functions $f_n : \mathcal{I} \to V$. The sequence (f_n) converges uniformly to f if

$$\forall \varepsilon > 0 \; \exists n_0(\varepsilon) > 0 \; \forall n \geq n_0 \; \forall t \in \mathcal{I} : \|f_n(t) - f(t)\|_V \leq \varepsilon$$

We write shortly

$$f_n \rightrightarrows f \text{ on } \mathcal{I} \quad \text{as } n \to \infty.$$

It is well-known that if the functions f_n are all in $C(\mathcal{I}, V)$ then also f is continuous. Here uniform convergence is convergence in the max-norm, i.e.

$$f_n \rightrightarrows f \text{ on } \mathcal{I} \quad \Leftrightarrow \quad \max_{t \in \mathcal{I}} \|f_n(t) - f(t)\|_V \to 0 \quad \text{as } n \to \infty.$$

Theorem A.18 (Criterion for uniform convergence).
Let (f_n) be a sequence of functions $f_n : \mathcal{I} \to V$. Suppose the functions $f_n \in C(\mathcal{I}, V)$ are equicontinuous on $\mathcal{I}$. Then the pointwise convergence

$$f_n(t) \to f(t) \quad \text{as } n \to \infty$$

for all $t \in \mathcal{I}$ implies the uniform convergence

$$f_n \rightrightarrows f \text{ on } \mathcal{I} \quad \text{as } n \to \infty$$

and the limit function $f : \mathcal{I} \to V$ is continuous.

Theorem A.19 (Arzelà-Ascoli).
A subset $H \subseteq C(\mathcal{I}, V)$ is relatively compact if and only if H is equicontinuous and for all $t \in \mathcal{I}$ the set

$$H(t) := \{f(t) \in V \mid f \in H\}$$

is relatively compact in V.

Bibliography

[ABCdF10] G. Alì, A. Bartel, M. Culpo, and C. de Falco. Analysis of a PDE Thermal Element Model for Electrothermal Circuit Simulation. In *Scientific Computing in Electrical Engineering (SCEE) 2008*, 2010.

[ABG05] G. Alì, A. Bartel, and M. Günther. Parabolic Differential-Algebraic Models in Electrical Network Design. *Multiscale Modeling & Simulation*, 4(3):813–838, 2005.

[ABG10] G. Alì, A. Bartel, and M. Günther. Existence and Uniqueness for an Elliptic PDAE Model of Integrated Circuits. *SIAM Journal on Applied Mathematics*, 70:1587–1610, 2010.

[ABGT03] G. Alì, A. Bartel, M. Günther, and C. Tischendorf. Elliptic Partial Differential-Algebraic Multiphysics Models in Electrical Network Design. *Mathematical Models and Methods in Applied Sciences*, 13(9):1261–1278, 2003.

[Ada75] R.A. Adams. *Sobolev Spaces*. Academic Press, New York, 1975.

[Bar04] A. Bartel. *Partial Differential-Algebraic Models in Chip Design - Thermal and Semiconductor Problems. Fortschritt-Berichte VDI*. Number 391 in Fortschritt-Berichte VDI, Reihe 20. VDI-Verlag, Düsseldorf, 2004. Dissertation.

[Bau12] S. Baumanns. *Coupled Electromagnetic Field/Circuit Simulation: Modeling and Numerical Analysis*. Dissertation, Universität zu Köln, 2012.

[BCP96] K.E. Brenan, S.L. Campbell, and L.R. Petzold. *Numerical Solution of Initial-Value Problems in Differential-Algebraic Equations*. SIAM, 1996.

[BG03] A. Ben-Israel and N.E. Greville. *Generalized Inverses: Theory and Applications*. Springer, New York, 2003.

[BJ11] M. Brunk and A. Jüngel. Self-heating in a coupled thermo-electric circuit-device model. *Journal of Computational Electronics*, 10:163–178, 2011.

[Bod07] M. Bodestedt. *Perturbation Analysis of Refined Models in Circuit Simulation*. Dissertation, Technische Universität Berlin, 2007.

[Cam87]   S.L. Campbell. A general form for solvable linear time varying singular systems of differential equations. *SIAM Journal on Mathematical Analysis*, 18(4):1101–1115, 1987.

[CC07]   V.F. Chistyakov and E.V. Chistyakova. Nonlocal theorems on existence of solutions of differential-algebraic equations of index 1. *Russian Mathematics (Iz Vuz)*, 51(1):71–76, 2007.

[CdF08]   M. Culpo and C. de Falco. A PDE thermal model for CHIP-level simulation including substrate heating effects. Preprint, DCU School of Math. Sciences, 2008.

[CDK87]   L.O. Chua, Ch.A. Desoer, and E.S. Kuh. *Linear and nonlinear Circuits*. McGraw-Hill Book Co., Singapore, 1987.

[CL55]   E. Coddington and N. Levinson. *Theory of Ordinary Differential Equations*. McGraw-Hill Book Co., 1955.

[CL75]   L.O. Chua and P.M. Lin. *Computer-Aided Analysis of Electronic Circuits: Algorithms & Computational Techniques*. Prentice-Hall, Englewood Cliffs, NJ, 1975.

[CM96a]   S.L. Campbell and W. Marszalek. ODE/DAE Integrators and MOL Problems. *Zeitschrift fuer Angewandte Mathematik und Mechanik (ZAMM)*, pages 251–254, 1996.

[CM96b]   S.L. Campbell and W. Marszalek. The index of an infinite dimensional implicit system. *Mathematical and Computer Modelling of Dynamical Systems*, 5:18 – 42, 1996.

[Cul09]   M. Culpo. *Numerical Algorithms for System Level Electro-Thermal Simulation*. Dissertation, Bergische Universität Wuppertal, 2009.

[Die85]   J. Dieudonné. *Grundzüge der modernen Analysis 1*. Deutscher Verlag der Wissenschaften, Berlin, 3rd edition, 1985.

[Die05]   R. Diestel. *Graph Theory*. Graduate Texts in Mathematics. Springer, 2005.

[DK84]   C.A. Desoer and E.S. Kuh. *Basic Circuit Theory*. International student edition. McGraw-Hill, 1984.

[Emm04]   E. Emmrich. *Gewöhnliche und Operator-Differentialgleichungen*. Vieweg, 2004.

[Est00]   D. Estévez-Schwarz. *Consistent initialization for index-2 differential algebraic equations and its application to circuit simulation*. Dissertation, Humboldt-Universität zu Berlin, 2000.

[ET00]   D. Estévez-Schwarz and C. Tischendorf. Structural analysis of electrical circuits and consequences for MNA. *International Journal of Circuit Theory and Applications*, 28:131–162, 2000.

[Eva08]   L.C. Evans. *Partial Differential Equations*. AMS, 2008.

[FP86]   A. Favini and P. Plazzi. Some results concerning the abstract degenerate nonlinear equation $D_t Mu(t) + Lu(t) = f(t, Ku(t))$. *Circuits, Systems, and Signal Processing*, 5(2):261–274, 1986.

[FR99]   A. Favini and A.G. Rutkas. Existence and uniqueness of solutions of some abstract degenerate nonlinear equations. *Differential and Integral Equations*, 12(3):373–394, 1999.

[FY99]   A. Favini and A. Yagi. *Degenerate differential equations in Banach spaces*. Marcel Dekker, New York, 1999.

[Gaj85]   H. Gajewski. On Existence, Uniqueness and Asymptotic Behavior of Solutions of the Basic Equations for Carrier Transport in Semiconductors. *ZAMM - Journal of Applied Mathematics and Mechanics / Zeitschrift für Angewandte Mathematik und Mechanik*, 65(2):101–108, 1985.

[Gaj93]   H. Gajewski. Analysis und Numerik von Ladungstransport in Halbleitern. *GAMM Mitteilungen*, 16:35–57, 1993.

[GG86]   H. Gajewski and K. Gröger. On the basic equations for carrier transport in semiconductors. *Journal of Mathematical Analysis and Applications*, 113:12–35, 1986.

[GGZ74]   H. Gajewski, K. Gröger, and K. Zacharias. *Nichtlineare Operatorgleichungen und Operatordifferentialgleichungen*. Akademie-Verlag Berlin, 1974.

[GM86]   E. Griepentrog and R. März. *Differential-Algebraic Equations and Their Numerical Treatment*. Teubner Verlagsgesellschaft, Leipzig, 1986.

[GP83]   C.W. Gear and L.R. Petzold. Differential/algebraic systems and matrix pencils. In B. Kågström and A. Ruhe, editors, *Matrix Pencils*, volume 973 of *Lecture Notes in Mathematics*, pages 75–89. Springer, Berlin/Heidelberg, 1983.

[Gün00]   M. Günther. Semidiscretization may act like a deregularization. *Mathematics and Computers in Simulation*, 53:293–301, 2000.

[Gün01]   M. Günther. *Partielle differential-algebraische Systeme in der numerischen Zeitbereichsanalyse elektrischer Schaltungen*. Number 343 in Fortschritt-Berichte VDI, Reihe 20. VDI Verlag, Düsseldorf, 2001. Habilitation.

[HLR89]   E. Hairer, C. Lubich, and M. Roche. *The numerical solutions of differential-algebraic systems by Runge-Kutta methods.* Springer-Verlag, Berlin, 1989.

[HM04]   I. Higueras and R. März. Differential algebraic equations with properly stated leading terms. *Computers & Mathematics with Applications*, 48:215–235, 2004.

[HNW02]   E. Hairer, S.P. Nørsett, and G. Wanner. *Solving Ordinary Differential Equations II: Stiff and Differential-Algebraic Problems.* Springer series in Computational Mathematics. Springer, Berlin, 2nd edition, 2002.

[HRB75]   C. Ho, A. Ruehli, and P. Brennan. The modified nodal approach to network analysis. *IEEE Transactions on Circuits and Systems*, 22(6):504–509, 1975.

[HW05]   H. Hoeber and A. Wachter. *Compendium of Theoretical Physics.* 2005.

[Jan13]   L. Jansen. Higher Index DAEs: Unique Solvability and Convergence of Implicit and Half-Explicit Multistep Methods. In preparation, 2013.

[Jit78]   K. Jittorntrum. An implicit function theorem. *Journal of Optimization Theory and Applications*, 25:575–577, 1978.

[JMT12]   L. Jansen, M. Matthes, and C. Tischendorf. Global Unique Solvability of Nonlinear Index-1 DAEs with Monotonicity Properties. Preprint, In preparation, 2012.

[Jos07]   J. Jost. *Partial Differential Equations.* Springer, 2007.

[Kam60]   E. Kamke. *Das Lebesgue-Stieltjes Integral.* Teubner Verlagsgesellschaft, Leipzig, 1960.

[KM94]   P. Kunkel and V. Mehrmann. Canonical forms for linear differential-algebraic equations with variable coefficients. *Journal of computational and applied mathematics*, 56(3):225–251, 1994.

[KM06]   P. Kunkel and V. Mehrmann. *Differential-Algebraic Equations: Analysis and Numerical Solution.* EMS Publishing House, Zürich, Switzerland, 2006.

[KP03]   S.G. Krantz and H.R. Parks. *The Implicit Function Theorem, History, Theory and Applications.* Birkhäuser, 2003.

[Kum80]   S. Kumagai. An implicit function theorem: Comment. *Journal of Optimization Theory and Applications*, 31:285–288, 1980.

[Lio69]   J. Lions. *Quelques méthodes de résolution des problèmes aux limites non linéaires.* Dunod Gauthier-Villars, 1969.

[LMT01]   R. Lamour, R. März, and C. Tischendorf. PDAEs and Further Mixed Systems as Abstract Differential Algebraic Systems. Preprint, Humboldt-Universität zu Berlin, 2001.

[LMT13]   R. Lamour, R. März, and C. Tischendorf. *Differential Algebraic Equations: A Projector Based Analysis.* Springer (in print), 2013.

[LSEL99]   W. Lucht, K. Strehmel, and C. Eichler-Liebenow. Indexes and Special Discretization Methods for Linear Partial Differential Algebraic Equations. *BIT Numerical Mathematics*, 39(3):484–512, 1999.

[Mar86]   P.A. Markowich. *The Stationary Semiconductor Devices.* Springer, Vienna, 1986.

[Mär03]   R. März. Differential-algebraic systems with properly stated leading term and MNA equations. In K. Anstreich, R. Bulirsch, A. Gilg, and P. Rentrop, editors, *Modelling, Simulation and Optimization of Integrated Circuits*, pages 135–151. Birkhäuser, 2003.

[MB00]   W.S. Martinson and P.I. Barton. A Differentiation Index for Partial Differential-Algebraic Equations. *SIAM Journal on Scientific Computing*, 21:2295–2315, 2000.

[Meh12]   V. Mehrmann. Index concepts for differential-algebraic equations. Preprint, Technische Universität Berlin, 2012.

[Moc83]   M.S. Mock. *Analysis of Mathematical Models of Semiconductor Devices.* Boole-Press, Dublin, 1983.

[MT11]   M. Matthes and C. Tischendorf. Convergence analysis of a partial differential algebraic system from coupling a semiconductor model to a circuit model. *Applied Numerical Mathematics*, 61(3):382–394, 2011.

[MT12]   M. Matthes and C. Tischendorf. Private communication, 2012.

[OR70]   J.M. Ortega and W.C. Rheinboldt. *Iterative solution of nonlinear equations in several variables.* Academic Press, 1970.

[RA05]   J. Rang and L. Angermann. Perturbation index of linear partial differential-algebraic equations. *Applied Numerical Mathematics*, 53(2-4):437–456, 2005.

[Rei91]   S. Reich. On an existence and uniqueness theory for nonlinear differential-algebraic equations. *Circuits, Systems, and Signal Processing*, 10:343–359, 1991.

[Rei06]   T. Reis. *Systems Theoretic Aspects of PDAEs and Applications to Electrical Circuits*. Dissertation, Technische Universität Kaiserslautern, 2006.

[Rei07]   T. Reis. Consistent initialization and perturbation analysis for abstract differential-algebraic equations. *Mathematics of Control, Signals, and Systems*, 19(3):255–281, 2007.

[Ria08]   R. Riaza. *Differential-Algebraic Systems: Analytical Aspects and Circuit Applications*. World Scientific, 2008.

[RK04]   A.G. Rutkas and I.G. Khudoshin. Global solvability of one degenerate semilinear differential operator equation. *Nonlinear Oscillations*, 7:403–417, 2004.

[Rou05]   T. Roubíček. *Nonlinear Partial Differential Equations with Applications*. Birkhäuser Verlag, Basel Boston Berlin, 2005.

[RT05]   T. Reis and C. Tischendorf. Frequency domain methods and decoupling of linear infinite dimensional differential algebraic systems. *Journal of Evolution Equations*, 5(3):357–385, 2005.

[Rut07]   A.G. Rutkas. Spectral methods for studying degenerate differential-operator equations. I. *Journal of Mathematical Sciences*, 144:4246–4263, 2007.

[Rut08]   A.G. Rutkas. Solvability of semilinear differential equations with singularity. *Ukrainian Mathematical Journal*, 60(2):262–276, 2008.

[RV03]   A.G. Rutkas and L.A. Vlasenko. Existence, uniqueness and continuous dependence for implicit semilinear functional differential equations. *Nonlinear Analysis*, 55:125–139, 2003.

[Saa03]   Y. Saad. *Iterative Methods for Sparse Linear Systems*. Society for Industrial and Applied Mathematics, Philadelphia, PA, USA, 2nd edition, 2003.

[Sch02]   S. Schulz. Ein PDAE-Netzwerkmodell als Abstraktes Differential-Algebraisches System. Diplomarbeit, Humboldt-Universität zu Berlin, 2002.

[Sch11]   S. Schöps. *Multiscale Modeling and Multirate Time-Integration of Field/Circuit Coupled Problems*. Dissertation, Bergische Universität Wuppertal, 2011.

[Sel84]   S. Selberherr. *Analysis and Simulation of Semiconductor Devices*. Springer, Vienna, 1984.

[Sho97]   R.E. Showalter. *Monotone Operators in Banach Spaces and Nonlinear Partial Differential Equations*. American Mathematical Society, 1997.

[Sim56]   K. Simonyi. *Theoretische Elektrotechnik.* Deutscher Verlag der Wissenschaften, Berlin, 1956.

[Sim98]   B. Simeon. DAE's and PDE's in elastic multibody systems. *Numer. Algorithms*, 19:235–246, 1998.

[Sim00]   B. Simeon. *Numerische Simulation gekoppelter Systeme von partiellen und differential-algebraischen Gleichungen in der Mehrkörperdynamik.* Number 325 in Fortschritt-Berichte VDI, Reihe 20. VDI-Verlag, Düsseldorf, 2000. Habilitation.

[ST05]   M. Selva Soto and C. Tischendorf. Numerical Analysis of DAEs from coupled circuit and semiconductor simulation. *Applied Numerical Mathematics*, 53(2-4):471–488, 2005.

[Tis96]   C. Tischendorf. *Solution of index-2 differential algebraic equations and its application in circuit simulation.* Dissertation, Humboldt-Universität zu Berlin, 1996.

[Tis99]   C. Tischendorf. Topological index calculation of DAEs in circuit simulation. *Surveys on Mathematics for Industry*, 8(3-4):187–199, 1999.

[Tis04]   C. Tischendorf. Coupled Systems of Differential Algebraic and Partial Differential Equations in Circuit and Device Simulation. Humboldt-Universität zu Berlin, Habilitation, 2004.

[TS02]   U. Tietze and C. Schenk. *Halbleiter-Schaltungstechnik.* Springer, Berlin Heidelberg, 2002.

[Voi06]   S. Voigtmann. *General Linear Methods for Integrated Circuit Design.* Dissertation, Humboldt-Universität zu Berlin, 2006.

[Wer05]   D. Werner. *Funktionalanalysis.* Springer-Verlag, Berlin Heidelberg, 2005.

[Zei86]   E. Zeidler. *Nonlinear Functional Analysis and its Applications I. Fixed Point Theorems.* Springer-Verlag, New York, 1986.

[Zei90a]   E. Zeidler. *Nonlinear Functional Analysis and its Applications II/A. Linear Monotone Operators.* Springer Verlag, New York, 1990.

[Zei90b]   E. Zeidler. *Nonlinear Functional Analysis and its Applications II/B. Nonlinear Monotone Operators.* Springer Verlag, New York, 1990.

Notation

Abbreviation

ODE	ordinary differential equation
PDE	partial differential equation
DAE	differential-algebraic equation
ADAE	abstract differential-algebraic equation
PDAE	partial differential-algebraic equation
MNA	Modified Nodal Analysis
KCL	Kirchhoff's current law
KVL	Kirchhoff's voltage law
LI-cutset	cutset of inductors and current sources
CV-loop	loop of capacitors and voltage sources
w.r.t.	with respect to
f.a.a.	for almost all
e.g.	exempli gratia
cf.	confer
i.e.	id est

General

$\exists$	there exists
$\forall$	for all
$\mathbb{N}$	natural numbers
$\mathbb{R}$	real numbers
$\mathbb{R}^n$	real n-dimensional space
$\mathbb{C}$	complex numbers
$I_n \in \mathbb{R}^{n \times n}$	identity matrix
$A \in \mathbb{R}^{n \times m}$	real matrix with n rows and m columns
$A^\top \in \mathbb{R}^{m \times n}$	transpose of A
$\text{diag}\{a_1, \ldots, a_n\}$	diagonal matrix with entries a_i, $i = 1, \ldots, n$
$\ker A$	kernel of A
$\text{im}\, A$	image of A
$\text{rk}\, A$	rank of A
$\|\cdot\|$	Euclidean norm on $\mathbb{R}^n$
$(\cdot | \cdot)$	canonical scalar product on $\mathbb{R}^n$
$\|A\|_*$	matrix norm of A corresponding to vector norm $\|\cdot\|$

$\mathcal{I}$	time interval
M, N	sets
$x \in M$	x is an element of M
$x \notin M$	x is not an element of M
$M \subseteq N$	M is a subset of N
$M \cup N$	union of M and N
$M \cap N$	intersection of M and N
$M \oplus N$	direct sum of M and N
$M \times N$	direct product of M and N (product set)
$\dim X$	dimension of a vector space X
$Y^\perp$	orthogonal complement of the subspace $Y \subseteq X$
$\mathrm{span}\,\{M\}$	span of vectors in the set $M \subseteq X$
$f : M \to N$	map from M to N
$\mathrm{dom}(()f)$	domain of f
$f\|_U$	restriction of $f : M \to N$ to $U \subseteq M$
$f'(\cdot), \frac{\mathrm{d}}{\mathrm{d}t}f(t)$	derivative of $f : \mathcal{I} \to N$
$f_x(x, y)$	Jacobian of f w.r.t. x
$\partial_i f$	partial derivative of f in direction i
∇f	gradient of f
$\nabla \cdot f$	divergence of f
$|\Omega| = \mathrm{vol}_n(\Omega)$	measure of the nonempty open bounded set $\Omega \subseteq \mathbb{R}^n$
$\partial\Omega = \Gamma$	boundary of the set Ω
$\overline{\Omega}$	closure of the set Ω
$\overset{\circ}{\Omega}$	interior of the set Ω
$\mathrm{vol}_{n-1}(\Gamma)$	$(n-1)$-dimensional measure of Γ
$C(\Omega)$	space of continuous functions $f : \Omega \to \mathbb{R}$
$C(\overline{\Omega})$	all $f \in C(\Omega)$ that can be extended continuously to the boundary
$C^k(\Omega)$	space of k-times continuously differentiable functions $f : \Omega \to \mathbb{R}$
$C^\infty(\Omega)$	space of infinitely continuously differentiable functions $f : \Omega \to \mathbb{R}$
$C_0^\infty(\Omega)$	space of all $f \in C^\infty(\Omega)$ with compact support
$L_p(\Omega)$	space of p-integrable functions $f : \Omega \to \mathbb{R}$ $(p \geq 1)$
$L_\infty(\Omega)$	space of essentially bounded functions
$L_{1,\mathrm{loc}}(\Omega)$	space of locally integrable functions
$H^1(\Omega)$	Sobolev space of all functions $f \in L_2(\Omega)$ with generalized derivatives $\partial_i f \in L_2(\Omega)$
$H_0^1(\Omega)$	closure of $C_0^\infty(\Omega)$ in $H^1(\Omega)$
$H^{-1}(\Omega)$	space of linear continuous functionals on $H_0^1(\Omega)$
V, W	Banach spaces
$L(V, W)$	set of linear, continuous maps $V \to W$
$L(V)$	$= L(V, V)$
V^*	dual space of V

$\|\cdot\|_V$	norm on V
$\langle v^*, v \rangle_V$	$= v^*(v)$ with $v^* \in V^*$, $v \in V$
$\mathcal{A}^* \in L(W^*, V^*)$	dual operator of $\mathcal{A} \in L(V, W)$
H	Hilbert space
$(\cdot | \cdot)_H$	scalar product on H
$\mathcal{A}^{*H} \in L(H)$	adjoint operator of $\mathcal{A} \in L(H)$ in the Hilbert sense
$C(U, W)$	space of continuous functions $f : U \to W$, $U \subseteq V$
$C^1(U, W)$	space of continuously differentiable functions $f : U \to W$
$L_p(\mathcal{I}, V)$	space of p-integrable functions $f : \mathcal{I} \to V$
$v_n \to v$	(strong) convergence of the sequence $(v_n) \subseteq V$ to v as $n \to \infty$ in the norm $\|\cdot\|_V$
$v_n \rightharpoonup v$	weak convergence the sequence $(v_n) \subseteq V$ to v as $n \to \infty$, i.e. for all $f \in V^*$: $\langle f, v_n \rangle_V \to \langle f, v \rangle_V$
$v_n \rightrightarrows v$	convergence of the sequence $(v_n) \subseteq C(\mathcal{I}, V)$ to v as $n \to \infty$
$\overline{\lim}_{n \to \infty}$	limes superior
$\underline{\lim}_{n \to \infty}$	limes inferior

Modified Nodal Analysis

q_C	constitutive relation for capacitors (charge)
g_R	constitutive relation for resistors (conductance)
ϕ_L	constitutive relation for inductors (flux)
i_s	source term for current sources
v_s	source term for voltage sources
$e \in \mathbb{R}^{n_e}$	node potentials (dimension n_e)
$j_L \in \mathbb{R}^{n_L}$	currents through inductors (dimension n_L)
$j_V \in \mathbb{R}^{n_V}$	currents through voltage sources (dimension n_V)
$A_X \in \mathbb{R}^{n_e \times n_X}$	incidence matrix of branch type $X \in \{C, R, L, V, I\}$
$\mathrm{q}_C \in \mathbb{R}^{n_e \times k_{\overline{C}}}$	columns are orthonormal basis of ker $A_C^\top$
$\mathrm{p}_C \in \mathbb{R}^{n_e \times k_C}$	columns are orthonormal basis of $(\ker A_C^\top)^\perp$
$\mathrm{q}_{CV} \in \mathbb{R}^{k_{\overline{C}} \times k_{\overline{C}V}}$	columns are orthonormal basis of ker $A_V^\top \mathrm{q}_C$
$\mathrm{p}_{CV} \in \mathbb{R}^{k_{\overline{C}} \times k_{\overline{C}V}}$	columns are orthonormal basis of $(\ker A_V^\top \mathrm{q}_C)^\perp$
e_C	$= \mathrm{p}_C^\top e \in \mathbb{R}^{k_C}$
$e_{\overline{C}}$	$= \mathrm{q}_C^\top e \in \mathbb{R}^{k_{\overline{C}}}$
$e_{\overline{C}V}$	$= \mathrm{p}_{CV}^\top e_{\overline{C}} \in \mathbb{R}^{k_{\overline{C}V}}$
$e_{\overline{C}V}$	$= \mathrm{q}_{CV}^\top e_{\overline{C}} \in \mathbb{R}^{k_{\overline{C}V}}$
$A_{\overline{C}X}$	$= \mathrm{q}_C^\top A_X$ for $X \in \{R, L, V, I\}$
$A_{\overline{C}VX}$	$= \mathrm{q}_{CV}^\top \mathrm{q}_C^\top A_X$ for $X \in \{R, L, I\}$

Index

A
ADAE 5
ADAE Index 14
Applied potential 114

B
Banach space 5
   reflexive 156
   separable 156

C
Carathéodory conditions 154
Convergence
   in the norm 6
   uniform 162
   weak 6
Coupled systems 19, 101
   elliptic prototype 102
   parabolic prototype 119

D
Degenerate differential equation 18
Derivative
   classical 8
   Fréchet 13
   generalized 158
   weak 8
Differential-algebraic equation
   obvious constraint set 31
   semi-linear 11, 30
   with monotonicity properties 30
Differentiation Index 12
Dual space 5

E
Electrostatic potential 113
Equicontinuous set 162
Euclidean scalar product 6
Evolution equation 155
Evolution triple 155

F
Function
   cutoff 44, 143
   integrable 7
   Lipschitz continuous 24
   locally integrable 7
   locally Lipschitz continuous 24
   locally strongly monotone 25
   monotone decreasing 142
   strongly monotone 24

G
Galerkin approach 18
Galerkin scheme 160
Graph
   branch 44
   cutset 44
   incidence matrix 44, 113, 139
   loop 44
   node 44

H
Heat equation 138
Hilbert space 6

I
Inequality

classical 70
Hölder 157
Poincaré-Friedrich 116
Integration by parts formula 58, 159

K
Kronecker Index 12

L
Laplace equation 113
Lebesgue space 7, 157

M
Method of lines 13
Modified Nodal Analysis 42
    basic elements 42
    branch currents 44
    constitutive relations 42
    CV-loop 46
    decoupled equations 47
    equations 45
    global passivity 43
    Kirchhoff's current law 44
    Kirchhoff's voltage law 44
    LI-cutset 46
    local passivity 43
    node potentials 45
    short circuit 45
    topological index 1 conditions 46
Monotonicity trick 86, 111, 130
Moore-Penrose inverse 50

N
Norm
    Euclidean norm 6
    operator norm 6

O
Operator
    adjoint 61
    dual 56
    hemicontinuous 161
    Lipschitz continuous 6

monotone 161
    self-adjoint 61
    strongly monotone 161

P
Partial differential-algebraic equation . 5
Perturbation Index 12, 16
Polynomials 159
Product rule 64
Projector 153
    complementary 153
    orthogonal 153
Properly stated leading term 14, 30

R
Relatively compact set 162

S
Semigroup approach 18
Sobolev space 7
Solution function 28
Strangeness Index 12

T
Theorem
    Arzelà-Ascoli 163
    Browder Minty (on $\mathbb{R}^n$) 27
    Browder-Minty 161
    Carathéodory 154
    Hadamard 28
    Implicit Function Theorem 28
    Riesz Representation Theorem . . 156
    Zarantonello 27
Topological sum 153
Tractability Index 12